D0686818

INTRODUCTION TO THE
PHYSICS OF STELLAR INTERIORS

ASTROPHYSICS AND SPACE SCIENCE LIBRARY

A SERIES OF BOOKS ON THE RECENT DEVELOPMENTS
OF SPACE SCIENCE AND OF GENERAL GEOPHYSICS AND ASTROPHYSICS
PUBLISHED IN CONNECTION WITH THE JOURNAL
SPACE SCIENCE REVIEWS

VOLUME 34

INTRODUCTION TO THE PHYSICS OF STELLAR INTERIORS

by

V. KOURGANOFF
University of Paris-Sud, Orsay

D. REIDEL PUBLISHING COMPANY
DORDRECHT-HOLLAND / BOSTON-U.S.A.

INTRODUCTION À LA PHYSIQUE DES INTÉRIEURS STELLAIRES

First published in 1969 by Dunod, Paris

Translated from the French by Janet Rountree Lesh

Library of Congress Catalog Card Number 72-86104

ISBN 90 277 0279 9

Published by D. Reidel Publishing Company,
P.O. Box 17, Dordrecht, Holland

Sold and distributed in the U.S.A., Canada, and Mexico
by D. Reidel Publishing Company, Inc.
306 Dartmouth Street, Boston,
Mass. 02116, U.S.A.

Printed in The Netherlands by D. Reidel, Dordrecht

PREFACE

All astrophysicists are acquainted with the fundamental works of S. Chandrasekhar [6] and M. Schwarzschild [1] concerning the *internal structure* of stars. Although both of these works accentuate the principal mathematical devices of the theory (and use, for this reason, notations that are rather perplexing for the non-specialist), the work of Schwarzschild is distinguished by care in demonstrating the *physical meaning* of the principal equations, while that of Chandrasekhar makes every effort *not to skip a single step* in the calculations. On the other hand, Schwarzschild, who considers his two introductory chapters as simple *reviews* of results which are already known, passes a bit rapidly over certain difficult arguments, and Chandrasekhar never goes far enough in the analysis of the physical mechanisms involved.

From another point of view, the excellent review articles published in the *Encyclopedia of Physics* [5] by M. H. Wrubel, P. Ledoux, and others, and those published in *Stars and Stellar Systems* [4] by H. Reeves, B. Strömgren, R. L. Sears and R. R. Brownlee, and others, are principally intended for research workers who are *already initiated* into the theory of internal structure. These monographs are on a level that is clearly too high for the *general physicist* who is approaching these astrophysical questions for the first time, and more particularly for the *post-graduate student.*

The many readers in this category require both the continuity of a Chandrasekhar (logical presentation of the arguments and the calculations) and the 'physical intuition' of a Schwarzschild or a Strömgren.

For this reason we have been led to reconsider this set of treatises and monographs from the pedagogical point of view, and to try to *clarify the principal physical concepts* used in 'formulating' the theory; for experience in teaching has shown us that these concepts are generally poorly understood by most beginners. The systematic use of *numerical* integration (in an approximation which makes it possible to follow the *details* of the calculations) makes the reader aware of the *orders of magnitude* of the parameters involved. This is all the more reasonable in that we are faced with a problem in which the data and the results are almost always defined by numerical *tables*, and not in terms of functions with known mathematical properties.

All this leads us, after a brief review of some classical results, to emphasize the derivation of the equation of mechanical equilibrium for the (gaseous) stellar material – which is often likened, on the basis of a purely formal analogy, to the equation of equilibrium for a liquid at rest.

Next we take as a *starting point*, a sort of working hypothesis, the density distribution $\varrho(r)$ (a function of the distance r from the center of the star) obtained by

Schwarzschild (at the end of his book) for a model of the Sun in an evolutionary state close to its present state. Then we dissect, so to speak, the *physical mechanism* by which the density distribution $\varrho(r)$ determines the mass, pressure, and temperature distributions in the interior of the star.

We then come to the problem of the energy equilibrium of the star, which leads us to introduce the principal nuclear reactions involved (the p-p chain and the C-N cycle). We specify the *composition* of the nuclei by appropriate schematic representations which enable the beginner to follow more easily the details of the reactions (which he is all too inclined to consider as a simple play on notation). The use of the recent monograph by Fowler *et al.* [2] enables us to specify the numerical values of the atomic weights and the cross-sections, on which the actual calculation of the energy output from nuclear reactions depends.

We pass somewhat rapidly over the 'tunnel' effect, which is very well treated in most of the classical works on quantum mechanics, but we emphasize the *specifically thermonuclear nature* of the opposition between the favourable effect of an increase in temperature on the overcoming of the electrostatic repulsion of the nuclei, and the unfavourable effect of an excessive increase in temperature on the phenomenon of 'fusion' as such.

We likewise devote special attention to finding the physical meaning of certain ideas which are generally – and wrongly – considered to be 'obvious', such as the *convergence towards an equilibrium state* of a set of cyclical reactions, or the '*mean duration of cycle*'.

After a brief review of the *empirical representation of the energy outputs* ε_{pp} and ε_{CN} as a function of various parameters and of the temperature T, we proceed to a 'final test' which shows how taking into account the energy production and the opacity* of the stellar material 'justifies' (and consequently determines) the distribution $\varrho(r)$ used as a starting point.

We conclude with a presentation of the great discovery of M. Schwarzschild: the necessity of considering the *evolution* of stellar models if one wants to understand the structure of the static models.

A section of the last chapter summarizes the essentials of the fundamental mathematical concepts used in the construction of models, leaving out everything that makes the more 'advanced' accounts cumbersome for a beginner.

Finally, since this book is intended not only for the general physicist but also – and especially – for the student, we give at the end of each chapter a large number of exercises, mostly drawn from recent original publications. These problems are given with 'answers'**, which enable the student to check himself; but no detailed solution is given, as this might make the student too passive. The solution of such exercises

* We assume that the reader is familiar with the theory of the mechanical effects of radiation (radiation pressure), with the notion of opacity, and with their application to the theory of stellar structure. These concepts are explained, in great detail and in a completely elementary fashion, in Chapter VII of our book *Introduction to the General Theory of Particle Transfer* [9].

** Except for Exercises 1′ and 2′, which are especially easy.

constitutes an excellent introduction to theoretical research for all young physicists, whatever their ultimate field of specialization.

It goes without saying that the present work, whose ambitions are purposely limited, constitutes only an *introduction* to the theory of internal structure and of thermonuclear reactions. The reader who wishes to pursue the subject can turn to the more 'advanced' works already mentioned above; and since the bibliography in these works generally stops around 1963 (except for Fowler's monograph), we place at his disposal a list of references to more recent publications (up to the end of 1971).

We are happy to take this opportunity to thank Dr L. Bottinelli, maître-assistant in Astrophysics at the University of Paris-Sud (Orsay), who drew up some of the questions and most of the 'answers' for the exercises, under our direction.

TABLE OF CONTENTS

PREFACE V

CHAPTER I. GENERAL CONSIDERATIONS CONCERNING THE ENERGY RADIATED BY STARS

1. The Energy Output and Its 'Spectral Composition' 1
2. The Observational Data 1
3. Generalities Concerning the Energy Sources 2

CHAPTER II. MECHANICAL EQUILIBRIUM: THE EQUILIBRIUM BETWEEN THE GRAVITATIONAL FORCE PER UNIT VOLUME AND THE GRADIENT OF THE TOTAL PRESSURE

1. Introduction 3
2. The Equilibrium between the Gradient of the Total Pressure and the Gravitational Force per Unit Volume 4
   2.1. Newton's Theorem. The Gravitational Force per Unit Volume 4
   2.2. The Force per Unit Volume Produced by the Pressure Gradient 8
   2.3. The Equation Expressing the Mechanical Equilibrium of $(\mathrm{d}V)$ 9
3. The Relation between M_r and the Density ϱ at a Distance r from the Center 10
4. The Expression for div $\mathbf{g}$ as a Function of the Local Density ϱ. Poisson's Equation 10
5. The Calculation of the Gas Pressure P_{gas}. The Concept of the Mean Mass μ of a Particle of the Mixture in Units of m_{H} (where m_{H} is the Mass in Grams of a 'Real' Microscopic Hydrogen Atom) 12
6. A Model of the Sun at 'Constant Density' $\varrho=\bar{\varrho}$ 15
7. The 'Homologous' Model. Expressions for P_c and T_c in Terms of M and R 18

Exercises 20

CHAPTER III. THE DETERMINATION OF THE INTERNAL STRUCTURE BY THE DENSITY DISTRIBUTION $\varrho(r)$

1. Introduction 24
2. The Determination of the Distribution of the Mass M_r Contained in a Sphere of Radius r 25

3. The Determination of the Distribution of the Total Pressure P as a Function of r 28
4. The Determination of the Distribution of the Temperature T as a Function of r 31
5. Summary. The Empirical Representation of the Functions $g(r')$, $\varrho(r')$, $P(r')$, and $T(r')$. The Polytropic Index n 32
6. The (Superficial) 'Convective Zone' of the Sun 35
Exercises 37

CHAPTER IV. ENERGY EQUILIBRIUM AND NUCLEAR REACTIONS

1. The Equation of Energy Equilibrium 44
2. The p-p Chain and the C-N Cycle 45
2.1. Introduction 45
2.2. An Explicit Schematic Representation of the Composition of Nuclei 45
2.3. The Details of the Reactions in the p-p Chain (Bethe, 1938) 45
2.4. The Details of the C-N Cycle (Bethe, 1938) 49
3. Calculation of the Energy ε. Generalities 52
3.1. Calculation of R_{12} for a Given Reaction 52
4. The 'Mean Lifetime' of a Given Nucleus with Respect to an Isolated Reaction (R) 57
4.1. Generalities 57
4.2. The Physical Meaning of $\tau_p(c)$ 59
4.3. $\tau_p(c)$ as the 'Mean Duration of an Isolated Reaction (R)' 60
4.4. $\tau_p(c)$ as an 'Exponential Decrement' 61
4.5. The Transition Probability p_{ca} per Reaction (R) 61
5. The Convergence of Cyclic Reactions to a Stationary ('Equilibrium') State 61
6. The 'Mean Duration of a Cycle'. The Calculation of the Energy ε when Cyclic Reactions Are Present 68
7. The Empirical Representation of ε_{pp} and ε_{CN} 70
8. Application to the Sun. The 'Final Test' 73
8.1. Review of the Principal Results 73
8.2. The Region in which ε Is Negligible 74
8.3. The 'Central' Region ($r'<0.40$), where L'_r and X Vary 76
8.4. The 'Final Test' 77
Exercises 78

CHAPTER V. EVOLUTIONARY MODELS. THE ACTUAL DETERMINATION OF STRUCTURE

1. Introduction 84
1.1. The Advantage of Studying 'Evolutionary Sequences' 85
1.2. The 'Fossilized' Composition 85
2. The Evolution of the Distributions $X(r)$ and $Y(r)$ 87
3. Discussion 89

4. The Mathematical Structure of the Problem. Principles of the Integration Methods 93
5. The Age of a Star 97
6. The Relations between P, T, L, R, and Parameters such as M, k_0, ε_0, and μ for 'Homologous' Models. The 'Mass-Luminosity' and 'Mass-Radius' Relations 98

CONCLUSION 101

SOLUTIONS FOR THE EXERCISES 104

BIBLIOGRAPHY 110

INDEX OF SUBJECTS 114

CHAPTER I

GENERAL CONSIDERATIONS CONCERNING THE ENERGY RADIATED BY STARS

1. The Energy Output and Its 'Spectral Composition'

The energy output of a star and its spectral composition depend on:

(a) The *physical conditions* prevailing in the star;

(b) The *chemical composition* of the star.

But stars are neither physically nor chemically homogeneous, and their inhomogeneity introduces the influence of:

(c) The *structure* of the star – that is, the variation of the physical and chemical conditions from one layer to the next, even if there is spherical symmetry (neglecting the rotation of the star).

Moreover, the physical and chemical conditions change, and their evolution introduces the general influence of:

(d) The *time factor*. This influence can be continuous and regular, or explosive; it can be connected with the stellar radiation itself, with pulsations, or with cataclysmic events.

2. The Observational Data

Our knowledge of the internal structure of stars and of the origin of their radiation rests on the following observational data:

(1) The apparent brightness and the distance give the (total bolometric) luminosity L of the star – that is, the total energy radiated in ergs per second.

(2) The study of star clusters gives luminosity ratios L/L' of several stars directly without requiring any knowledge of the distance of the cluster.

(3) Observations of the continuous spectrum and the laws of black-body radiation give the surface temperature T_s of the star.

(4) From the surface temperature T_s, we derive the surface radiation per unit area F_s, using Stefan's law for black-body radiation.

(5) Dividing L by F_s, we obtain the total surface area of the star and thence its radius R_0; the radius can also be measured by observing the eclipses of certain double stars whose orbital velocity is known.

(6) From the period of revolution of certain double stars for which we know the dimensions of the orbit, we can calculate the mass M of the star by using the newtonian interpretation of Kepler's third law.

(7) Comparison of the theoretical study of stellar atmospheres with observations of spectra gives the acceleration of gravity g at the surface of the star.

(8) Knowing g and R, we can obtain a determination of the mass M which is independent of the procedure indicated in (6).

(9) The theoretical quantitative analysis of the spectra of diffuse nebulae, interstellar matter, and stellar atmospheres gives the chemical composition of cosmical objects and of the outer layers of stars.

(10) The study of the relative motion of double stars gives the mass distribution in the interior of the stars.

All these data are compared with theoretical 'models' in which one assumes various values of the parameters (a), (b), (c), and (d) of Section 1, and conclusions are drawn concerning L, R, and M.

That combination of parameters is adopted, which gives the observed values of L, R, and M.

We generally discover that *only a very small choice* of models is possible.

3. Generalities Concerning the Energy Sources

In modern times, two principal processes can be regarded as explaining the radiation L of a star:

(1) *Gravitational contraction* ('free fall' of the outer layers towards the center), which transforms the potential energy of the system into the kinetic energy of thermal motion – that is, into heat which in turn gives rise to thermal radiation. This mechanism is important only in stars which are in the process of formation or in the process of collapse.

(2) *Thermonuclear reactions* ('adult' stars) that take place in the central regions, which are both the densest and hottest parts of the star (temperatures of the order of 10 million degrees). These nuclear reactions are accompanied by an energy release associated with the conversion of a fraction of the mass (not exceeding 1%) into energy.

Why do we speak of *thermo*-nuclear reactions?

Because the reacting nuclei, essentially protons, are all positively charged, and *repel* each other according to Coulomb's law. In order to overcome this repulsion (or, as we say in 'technical' language, in order to cross the 'potential barrier'), collisions at a large relative velocity are required.

In the laboratory, this large relative velocity can be imparted to projectiles launched at fixed targets by the electromagnetic forces of a particle 'accelerator'.

In a star, the relative velocity results from the natural thermal agitation of the particles. At a given temperature, all possible relative velocities can be found; but large velocities predominate at high temperatures. The mean kinetic energy of the particles is of the order of $\frac{3}{2}kT$. For $T = 20 \times 10^6$ K, with

$$k = 1.38 \times 10^{-16} \text{ c.g.s.},$$

we find a mean kinetic energy of the order of 3×10^{-9} erg or 2 *kiloelectron volts*.

To overcome the coulomb repulsion effectively, the relative velocity required for protons corresponds to 20 keV. With a velocity distribution whose mean kinetic energy is 2 keV, a *sufficiently large fraction* of the particles has the energy of 20 keV necessary to 'start up' the nuclear reactions.

CHAPTER II

MECHANICAL EQUILIBRIUM: THE EQUILIBRIUM BETWEEN THE GRAVITATIONAL FORCE PER UNIT VOLUME AND THE GRADIENT OF THE TOTAL PRESSURE

1. Introduction

Let us consider a spherical star of radius R, in equilibrium, stratified in homogeneous concentric layers, whose center is O. We introduce an outward-directed axis **Or** and a unit vector **r** on this axis, also outward-directed (see Figure 1).

Since the pressure, the temperature and the density increase towards the center, the gradients of all these quantities are inward-directed, and their projections on the axis **Or** are negative.

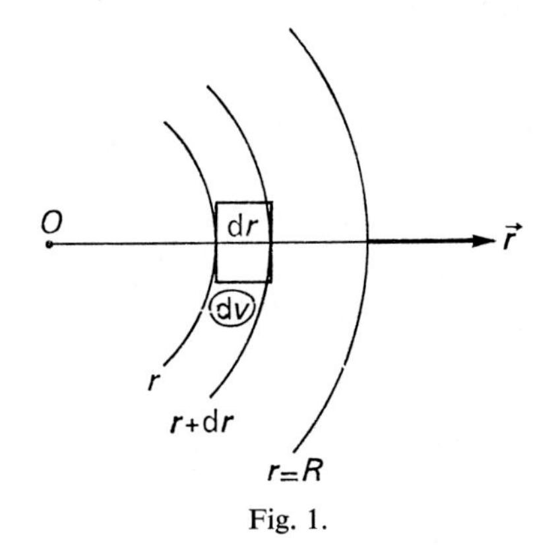

Fig. 1.

Consider a *strictly cylindrical* volume element $(\mathrm{d}V)$ of cross-section $\mathrm{d}A$ and length $\mathrm{d}r$, whose axis is **r** and whose generatrix is parallel to **r**. This element $(\mathrm{d}V)$ is subject to a gravitational interaction with the rest of the star '*at a distance*', and mechanical equilibrium is guaranteed by a *local* interaction with the neighbouring elements which exert a *pressure* on $(\mathrm{d}V)$. The pressures on the 'lateral' parts of the cylinder balance each other, by symmetry; but the 'bases', located at different levels (r) and $(r+\mathrm{d}r)$, experience different pressures because the densities and the temperatures at these levels are different. This gives a net outward-directed force which can balance the resultant of the gravitational forces attracting $(\mathrm{d}V)$ towards the center of the star.

In this elementary study, we naturally assume that there are no internal motions, not even stationary ones (*static* equilibrium).

Let us try to specify the conditions of this equilibrium. Note that in postulating the existence of an equilibrium state, we make it possible to determine the pressure gradient from the parameters governing the 'gravitational field' and the density, *without calling on* a detailed analysis of the *physical processes* responsible for the pressure (bombardment of the bases of $\mathrm{d}V$ by photons and by particles coming from the neighbouring elements).

2. The Equilibrium between the Gradient of the Total Pressure and the Gravitational Force per Unit Volume

The gravitational interaction between the star as a whole and the element $(\mathrm{d}V)$ located at a distance r from the center O of the star obeys the law of universal gravitation. Now, in potential theory one proves 'Newton's theorem', according to which the attraction exerted on an element $(\mathrm{d}V)$ inside the star is equivalent to that of the entire mass M_r of the part located inside the sphere of radius r centered on O; the effect of the layers (r, R) between r and R is zero. We shall begin with a proof of this fundamental theorem.

2.1. Newton's Theorem. The Gravitational Force per Unit Volume

In its modern version, the proof of this theorem rests on a transposition to gravitational forces of Gauss's theorem for *electrostatics*. (The historical order is obviously just the opposite.)

According to this theorem, given the field $\mathbf{E}$ created by a positive charge e_c located at C, and a closed surface (Σ_r) made up of elements $(\mathrm{d}A_r)$ oriented by means of outward-directed vectors $\mathbf{r}$ normal to (Σ_r), the 'emergent flux of $\mathbf{E}$ across (Σ_r)' is given by:

$$\int_{\Sigma_r} (\mathrm{d}A_r)\,(\mathbf{r}\cdot\mathbf{E}) = \left\{ \begin{array}{ll} 0 & \text{if } C \text{ is outside } (\Sigma_r) \\ \dfrac{4\pi}{k}\, e_c & \text{if } C \text{ is inside } (\Sigma_r) \end{array} \right\}. \tag{1}$$

In this expression, k is the dielectric constant in Coulomb's law of repulsion

$$\mathbf{E} = k^{-1} e_c d^{-2} \mathbf{r}\,, \tag{2}$$

which corresponds to the law of attraction by universal gravitation

$$\mathbf{g} = -\,G m_c d^{-2} \mathbf{r}\,. \tag{3}$$

G is the constant of gravitation (6.67×10^{-8} c.g.s.); m_c the mass in grams; d the distance in cm; and $\mathbf{g}$ the gravitational field – that is, the force exerted on unit mass, in dynes per gram.

When we transpose Gauss's theorem (1), keeping in mind the change from repulsion to attraction – that is, the fact that $\mathbf{g}$ is always directed towards the mass responsible for the gravitational field, we find that the 'emergent flux of the field $\mathbf{g}$' is given by:

$$\int_{\Sigma_r} (\mathrm{d}A_r)\,(\mathbf{r}\cdot\mathbf{g}) = \begin{cases} 0 & \text{if } m_c \text{ is outside } (\Sigma_r) \\ -4\pi G m_c & \text{if } m_c \text{ is inside } (\Sigma_r) \end{cases}. \tag{4}$$

Here is a proof of Equation (4) for those of our readers who may not be perfectly familiar with Equation (1).

We say that the element $(\mathrm{d}A_r)\,(\mathbf{r}\cdot\mathrm{g})$ of the emergent flux is given in absolute value by $Gm_c(\mathrm{d}\Omega)$, where $(\mathrm{d}\Omega)$ is the solid angle subtended by $(\mathrm{d}A_r)$ at the point C. For, let $\mathbf{\Omega}$ be the unit vector in the direction from $(\mathrm{d}A_r)$ to C. Then if C is outside (Σ_r), at C_1, we have (Figure 2):

$$\mathbf{g} = g\mathbf{\Omega} = +\,Gm_c d^{-2}\mathbf{\Omega}\,, \tag{5}$$

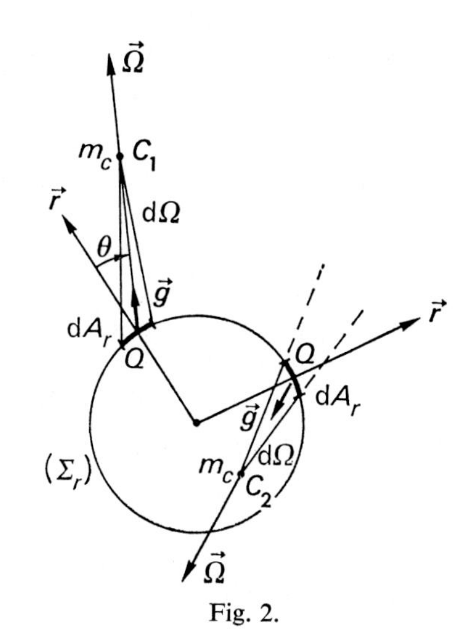

Fig. 2.

and, denoting the projection of $(\mathrm{d}A_r)$ on the plane normal to $\mathbf{\Omega}$ by $(\mathrm{d}A_\Omega)$:

$$\begin{aligned}(\mathrm{d}A_r)\,(\mathbf{r}\cdot\mathbf{g}) &= Gm_c(\mathrm{d}A_r)\,d^{-2}\cos\theta \\ &= Gm_c d^{-2}(\mathrm{d}A_\Omega) = Gm_c(\mathrm{d}\Omega).\end{aligned} \tag{6}$$

It is easily verified that this formula remains valid in absolute value when C is at C_2, inside (Σ_r).

Let us apply Equation (6) to the case in which m_c is located outside (Σ_r), dividing (Σ_r) into two parts (Σ') and (Σ'') by the curve of intersection with the cone tangent to (Σ_r) whose vertex is C_1 and whose aperture is the solid angle Ω.

Then (Figure 3) we have for the 'emergent flux of the field g':

$$\int_{\Sigma'} Gm_c\,\mathrm{d}\Omega - \int_{\Sigma''} Gm_c\,\mathrm{d}\Omega = Gm_c\Omega - Gm_c\Omega = 0, \tag{7}$$

for, as we see in Figure 3, all the elements of *emergent* flux are negative over (Σ'') and positive over (Σ').

On the other hand, for a point C_2 inside (Σ_r), all the angles between **r** and **g** are obtuse, and all the elements of emergent flux are negative. The sum of all the $(\mathrm{d}\Omega)$ is equal to 4π, and we have:

$$\int_{\Sigma_r} (\mathrm{d}A_r)\,(\mathbf{r}\cdot\mathbf{g}) = -\,Gm_c\cdot\int_{\Sigma_r}\mathrm{d}\Omega = -\,4\pi Gm_c. \tag{7'}$$

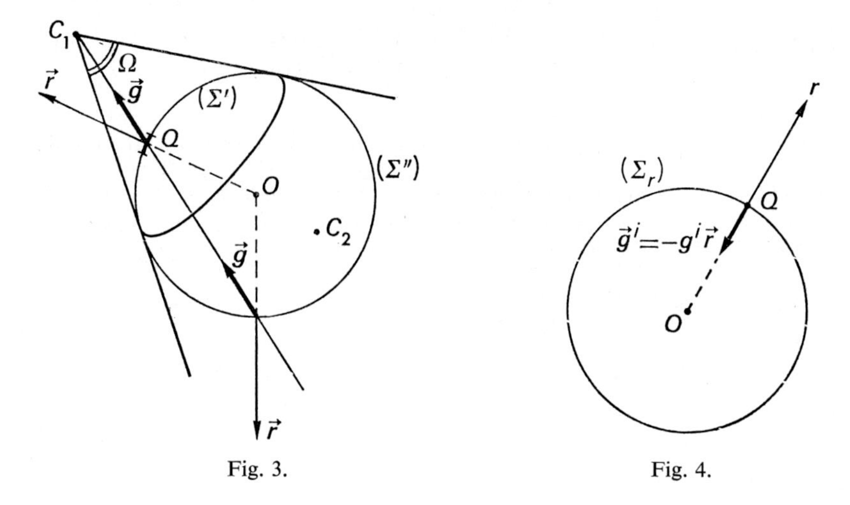

Fig. 3.

Fig. 4.

After this brief review of Gauss's theorem, let us return to Newton's theorem, which we intend to prove.

By symmetry, the field $\mathbf{g}^i$ produced at a given point at a distance r from the center O by all the mass points *inside* (Σ_r) must be *radial* under the hypothesis of *homogeneous* spherical layers; and its absolute value g^i must be the same at all points of (Σ_r) at the distance r from the center O (Figure 4).

Thus we have:

$$\mathbf{g}^i = -g^i\mathbf{r}; \qquad g^i = |\mathbf{g}^i|. \tag{8}$$

And for the (negative) 'emergent flux' of the field $\mathbf{g}^i$ over the sphere (Σ_r), we have:

$$\int_{\Sigma_r} (\mathrm{d}A_r)\,(\mathbf{r}\cdot\mathbf{g}^i) = -g^i \int_{\Sigma_r} (\mathrm{d}A_r) = -g^i 4\pi r^2. \tag{9}$$

But the sum of the quantities $(-4\pi G m_c)$ referring to the mass points 'enclosed' in (Σ_r) is equal to $(-4\pi G M_r)$. Thus we have, by Equations (4) and (9)

$$-g^i\cdot 4\pi r^2 = -4\pi G M_r, \tag{10}$$

that is,

$$g^i = G M_r r^{-2}. \tag{11}$$

Let us now consider the field $\mathbf{g}^e$ produced by the set of mass points *outside* the sphere (Σ_r) of radius r.

As a result of the hypothesis of stratification in homogeneous spherical layers, the field $\mathbf{g}^e$ (whatever it may be) for these external mass points must also have the same absolute value g^e at all points on the sphere (Σ_r) of radius r. Moreover, this field must by symmetry be radially directed (to see this, one has only to associate two by two the mass points which are symmetric with respect to $\mathbf{r}$). Since we do not know *a priori* in which sense $\mathbf{g}^e$ is directed, we shall set

$$\mathbf{g}^e = \varepsilon g^e \mathbf{r} \quad \text{with} \quad \varepsilon = \pm 1. \tag{12}$$

A calculation of the total flux, applying the definition of the 'emergent flux', gives, as in (9): $+\varepsilon g^e\, 4\pi r^2$. It follows, according to (4), that $g^e = 0$.

Qualitatively, this result can be physically interpreted as a balance (Figure 5) between the component $\mathbf{g}^e_+$ of the total field $\mathbf{g}^e$ which is created, at Q, by the diagonally hatched area (directed along $\mathbf{r}$ in the sense of $\mathbf{r}$), and the component $\mathbf{g}^e_-$ of $\mathbf{g}^e$ created by the vertically hatched area (directed along $-\mathbf{r}$). The component $\mathbf{g}^e_+$ is produced by a mass composed of a *small number* of elements *close* to Q, while the component $\mathbf{g}^e_-$ is produced by a mass composed of a *large number* of elements far from Q.

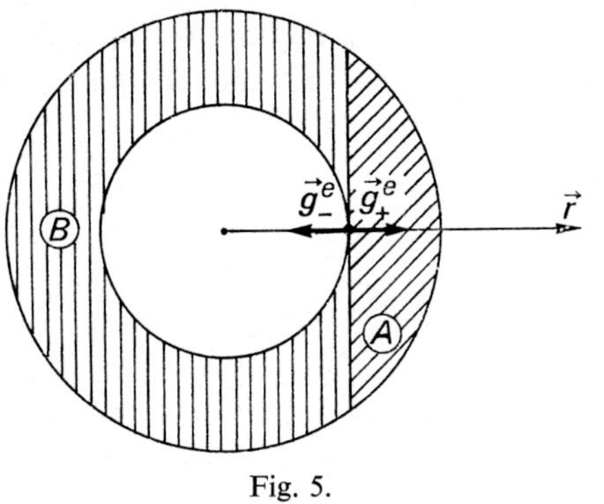

Fig. 5.

But the rigourous nature of this balance is one of those extraordinary simplifications with which Nature sometimes gratifies the scientist, and whose astonishing and almost miraculous quality was emphasized by Newton and Einstein.

We also note that one can, of course, prove Newton's theorem directly, without recourse to Gauss's theorem; but the method we have followed has the advantage of providing an immediate proof for the many readers who are well acquainted with classical electrostatics.

If we apply Newton's theorem at a point Q of the stellar interior, located at a distance r from the center, we have:

$$g = g^i = GM_r r^{-2}, \qquad \mathbf{g} = -g\mathbf{r}. \tag{13}$$

Then, denoting the density by $\varrho = \varrho(r)$ (a function of r alone, under the hypothesis of homogeneous layers) and the gravitational force acting on the mass of $(\mathrm{d}V)$ by $\mathrm{d}\mathbf{F}_{\text{grav}}$, we have

$$\mathrm{d}\mathbf{F}_{\text{grav}} = \mathbf{g}\cdot(\varrho\, \mathrm{d}V). \tag{14}$$

Hence, the gravitational force per unit volume is:

$$\mathrm{d}\mathbf{F}_{\text{grav}}/\mathrm{d}V = -\, GM_r r^{-2} \varrho \mathbf{r}. \tag{15}$$

This force per unit volume is directed towards the interior (more exactly, towards the center O), and its absolute value is:

$$|\mathrm{d}\mathbf{F}_{\text{grav}}/\mathrm{d}V| = GM_r r^{-2} \varrho. \tag{15'}$$

2.2. The Force per Unit Volume Produced by the Pressure Gradient

The exchange of material particles and of photons between the element $(\mathrm{d}V)$ and the *adjacent* elements of the hot gas forming the star produces dynamical effects – that is, forces which, when reduced to unit area, appear as *pressures.*

In order to clarify this idea of dynamical pressure, it is convenient to imagine the element $(\mathrm{d}V)$ as separated from the adjacent elements by a 'break' (a 'no man's land'), formed by an empty space across which the exchange of 'particles' (Figure 6) in the broad sense of the word (including photons) takes place.

The 'break' at the level (r) makes it possible to distinguish the surface (r^+) of $(\mathrm{d}V)$, facing the center O, from the surface (r^-) of the medium adjacent to (r^+) but outside

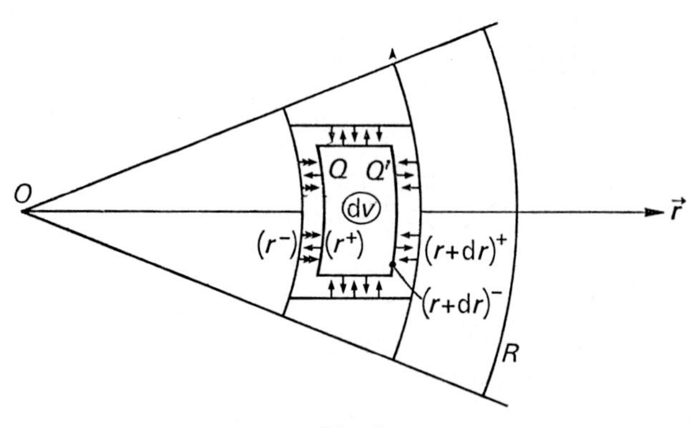

Fig. 6.

$(\mathrm{d}V)$. The same situation naturally occurs at the level $(r+\mathrm{d}r)$, where the surface of $(\mathrm{d}V)$ facing the 'outside' should be called $(r+\mathrm{d}r)^-$, in contrast to the surface $(r+\mathrm{d}r)^+$ of the adjacent medium outside $(\mathrm{d}V)$.

Consequently, we shall denote by $P^+(r)$ the pressure produced on the surface (r^+) of $(\mathrm{d}V)$ by the exchange of particles between (r^-) and (r^+). Likewise we shall denote by $P^-(r+\mathrm{d}r)$ the pressure produced on the surface $(r+\mathrm{d}r)^-$ of $(\mathrm{d}V)$ by the exchange of particles between $(r+\mathrm{d}r)^-$ and $(r+\mathrm{d}r)^+$.

As a specific example, let us begin by comparing $P^+(r)$ and $P^-(r)$, where the latter expression denotes the pressure produced on the surface $(r)^-$ of the adjacent medium by the exchange with $(\mathrm{d}V)$. Physically, $P^+(r)$ results from the addition of *two* 'components':

(1) an 'active' component produced by the impact of the particles (double arrows in Figure 6) coming from (r^-);

(2) a 'reactive' component produced by the ejection of particles (single arrows) from $(\mathrm{d}V)$ towards the center O – an ejection which tends to push $(\mathrm{d}V)$ towards the outside by 'reaction' (as with the ejection of gas from the nozzle of a rocket or a jet plane).

But the same particles appear, with reversed roles, in the sum of which $P^-(r)$ is composed: in this case, the single arrows correspond to the 'active component' and the double arrows to the 'reactive component'. The result is that $P^+(r)=P^-(r)$, and their common value may be denoted by $P(r)$.

This makes it possible to express the absolute value of the total pressure applied to the base of $(\mathrm{d}V)$ facing outwards at the level $(r+\mathrm{d}r)$ as $P(r+\mathrm{d}r)$ (see Figure 7).

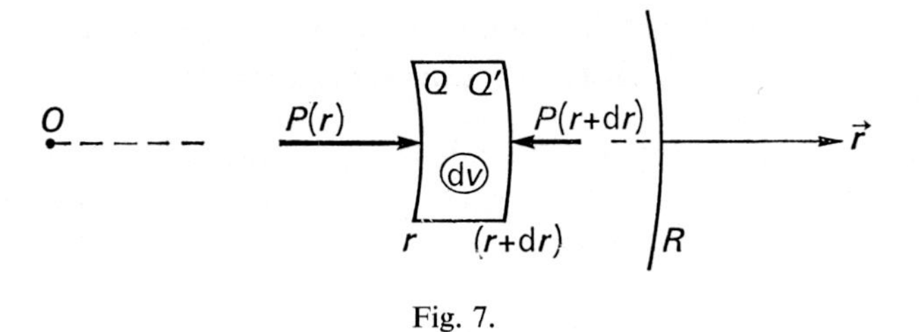

Fig. 7.

Thus the force $\mathrm{d}\mathbf{F}_{\text{press}}$ exerted on the whole of the cylindrical element $(\mathrm{d}V)$ is reduced to:

$$\begin{aligned}\mathrm{d}\mathbf{F}_{\text{press}} &= P(r)\,\mathrm{d}A\,\mathbf{r} - P(r+\mathrm{d}r)\,\mathrm{d}A\,\mathbf{r} \\ &= -\left[P(r+\mathrm{d}r) - P(r)\right]\mathrm{d}A\,\mathbf{r}\,,\end{aligned} \tag{16}$$

where $\mathrm{d}A$ is the common area of the two bases.

Since

$$\left[P(r+\mathrm{d}r) - P(r)\right] = \frac{\mathrm{d}P}{\mathrm{d}r}\,\mathrm{d}r\,,$$

we immediately deduce that

$$\frac{\mathrm{d}\mathbf{F}_{\text{press}}}{\mathrm{d}V} = -\frac{\mathrm{d}P}{\mathrm{d}r}\,\mathrm{d}r\,\mathrm{d}A\,\mathbf{r}/\mathrm{d}V = -\frac{\mathrm{d}P}{\mathrm{d}r}\,\mathbf{r} = -\,\mathbf{grad}\,P\,. \tag{17}$$

As P depends only on r, the vector $\mathbf{grad}P$ reduces to $(\mathrm{d}P/\mathrm{d}r)\mathbf{r}$.

2.3. The equation expressing the mechanical equilibrium of $(\mathrm{d}V)$

Since $\mathrm{d}P/\mathrm{d}r$ is negative, the absolute value of the force of *pressure* per unit volume is $(-\mathrm{d}P/\mathrm{d}r)$. It is outward-directed. Now, the force of *gravity* per unit volume is inward-directed, and according to Equation (15′) its absolute value is $GM_r\varrho/r^2$. For equilibrium, it is thus required that:

$$\boxed{-\,\mathrm{d}P/\mathrm{d}r = GM_r r^{-2}\varrho}\,. \tag{18}$$

This equation is equivalent to the vector equation

$$\mathbf{grad}\,P = \mathbf{g}\varrho\,, \tag{18′}$$

where the vector **g** given by (13) is *inward*-directed, as is the vector **grad**P (but the equilibrium is between the outward-directed $-$**grad**P and the inward-directed **g**ϱ!).

We have reasoned as if we knew that $\mathrm{d}P/\mathrm{d}r$ is negative, but in reality it is precisely from Equation (18) that we learn this fact – just as we learn, more generally, the value of $\mathrm{d}P/\mathrm{d}r$ (that is, $[-GM_r r^{-2}\varrho]$) without examining in detail the physical mechanism responsible for the existence of a pressure $P(r)$.

3. The Relation between M_r and the Density ϱ at a Distance r from the Center

Let us express M_r as a function of ϱ, the density at a distance r from the center. The mass of a layer $(\xi, \mathrm{d}\xi)$ inside (Σ_r), the sphere of radius r, is obtained by multiplying the volume $4\pi\xi^2\,\mathrm{d}\xi$ of that layer by ϱ.

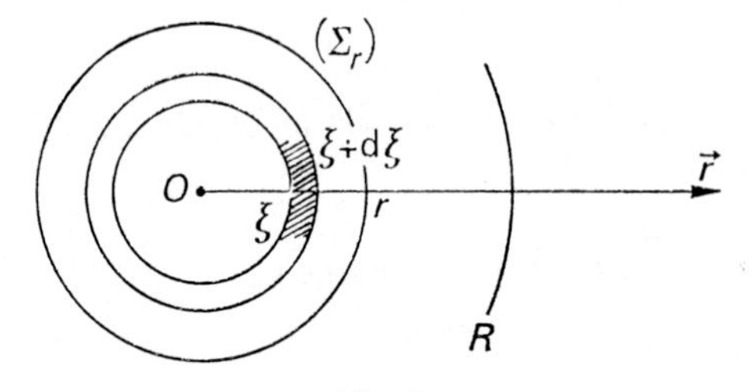

Fig. 8.

Thus we have (Figure 8):

$$M_r = \int_0^r \varrho 4\pi\xi^2\,\mathrm{d}\xi\,. \tag{19}$$

Similarly, the increase $(\mathrm{d}M_r)$ in M_r when r increases by $(\mathrm{d}r)$ is

$$\mathrm{d}M_r = 4\pi\varrho r^2\,\mathrm{d}r\,, \quad \text{whence} \quad \boxed{\mathrm{d}M_r/\mathrm{d}r = 4\pi\varrho r^2}\,. \tag{20}$$

Combining (20) with (13), we immediately obtain the following relation:

$$\mathrm{d}\,(r^2 g_r)/\mathrm{d}r = \mathrm{d}\,(GM_r)/\mathrm{d}r = G\,\mathrm{d}M_r/\mathrm{d}r = G4\pi\varrho r^2\,, \tag{21}$$

in which we have placed a subscript r on g to remind ourselves that we are dealing with the absolute value of the field **g** at the point Q at a distance r.

4. The Expression for div**g** as a Function of the Local Density ϱ. Poisson's Equation

Let us apply the definition of the divergence of a vector field at a point Q of the field. According to the 'physical' definition of the divergence, div**g** is merely the flux *density* of **g** (flux per unit volume) emerging from a small volume element containing Q.

Let us take as the volume element in question the spherical layer $(r, \mathrm{d}r)$ (Figure 9).

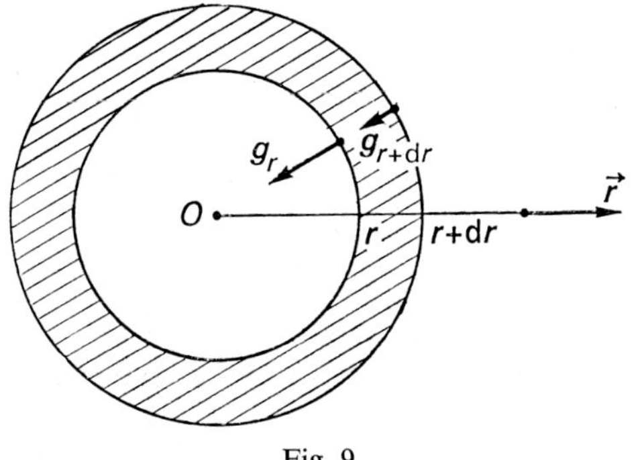

Fig. 9.

Dividing the 'flux of **g** emerging from $(r, \mathrm{d}r)$' by the volume $(r, \mathrm{d}r)$ and letting $\mathrm{d}r$ tend to zero, we have

$$\operatorname{div}\mathbf{g} = \lim \frac{4\pi r^2 g_r - 4\pi (r + \mathrm{d}r)^2 g_{r+\mathrm{d}r}}{4\pi r^2\, \mathrm{d}r} = -\frac{1}{r^2}\frac{\mathrm{d}(r^2 g_r)}{\mathrm{d}r}. \tag{22}$$

According to (21), this expression reduces to the elegant formula

$$\operatorname{div}\mathbf{g} = -4\pi\varrho G, \tag{23}$$

where $\operatorname{div}\mathbf{g}$ and ϱ refer to the same point Q at a distance r from the center.

This equation corresponds to the first of Maxwell's equations:

$$\operatorname{div}\mathbf{E} = 4\pi k^{-1}\varrho, \tag{23'}$$

where the electric field **E** has been replaced by the gravitational field **g**; the dielectric constant k has also been replaced by $(-G^{-1})$, and the charge density ϱ by the mass density.

On the other hand, since the field **g** is purely radial, its 'circulation' about any small curve traced on (Σ_r) is zero; thus, according to Stokes's theorem, the flux of **rot g** across the area enclosed by this curve will also be zero. This means that **rot g** is identically zero, and we can consider **g** as the gradient of some scalar field, writing:

$$\mathbf{g} = -\,\mathbf{grad}\,\varphi_g. \tag{24}$$

The function $\varphi_g(Q)$, which in the present case reduces to $\varphi_g(r)$, is called the 'gravitational potential' at the point Q.

Note that some authors set

$$\mathbf{g} = +\,\mathbf{grad}\,\varphi_g,$$

but only the formula with the minus sign preserves the analogy with the formulae of electrostatics, when k is changed to $(-1/G)$.

When G is expressed in this way as a function of the gravitational potential φ_g, the fundamental Equation (23) takes the form of 'Poisson's equation':

$$\boxed{\operatorname{div}(\mathbf{grad}\,\varphi_g) = \triangle\varphi_g = +\,4\pi G\varrho}\,, \tag{25}$$

where $\triangle\varphi_g$ is the 'Laplacian' of φ_g. This equation is analogous to Poisson's equation for electrostatics:

$$\triangle\varphi_e = -\,4\pi k^{-1}\varrho\,, \tag{25'}$$

where φ_e is the electrical potential and ϱ is the charge density.

5. The Calculation of the Gas Pressure P_{gas}. The Concept of the Mean Mass μ of a Particle of the Mixture, in Units of m_{H} (where m_{H} is the Mass in Grams of a 'Real' Microscopic Hydrogen Atom)

According to the perfect gas law, in the especially simple form suggested by the kinetic theory of gases, we have

$$P_{\text{gas}} = nkT\,, \tag{26}$$

where n is the *number* density (number of particles per cm^3 of the mixture), T is the absolute temperature, and k is a universal constant (Boltzmann's constant) whose numerical value will be given below.

According to this formula, P_{gas} does not depend 'fundamentally' on the *nature* of the particles, nor, in particular, on the *mass* in grams m_g of each of them.

However, the nature of the particles comes in (through their mass m_g) as soon as we seek the pressure P_{gas} corresponding to a given value of the density ϱ of the mixture – or, to put it another way, as soon as we calculate the ratio P_{gas}/ϱ.

In fact, the mean mass in grams $\bar{m}_g$ of each particle of the mixture is obtained by dividing the mass ϱ of one cm^3 of the gas by the number n of particles contained in this cm^3:

$$\bar{m}_g = \varrho/n\,. \tag{27}$$

Then let $m_{\text{H}} = 1.66 \times 10^{-24}$ grams, the mass of a hydrogen atom (actually, the 'mass of unit atomic weight' – but in the present approximation, setting this mass equal to that of a 'real' hydrogen atom is acceptable and makes possible a more concrete description). In order to evaluate $\bar{m}_g$ in units of m_{H}, we have only to set

$$\boxed{\mu = \bar{m}_g/m_{\text{H}}}\,. \tag{28}$$

Then we have

$$n = \varrho/\mu m_{\text{H}}\,, \tag{29}$$

and Equation (26) takes the form

$$P_{\text{gas}} = \frac{\varrho}{\mu} RT, \tag{30}$$

where R is given by

$$R = k/m_{\text{H}}. \tag{31}$$

In order to find the value of the constant k, we have only to apply Equation (26) to 1 gram of atomic hydrogen under 'standard' conditions of temperature

$$(T^0 = 0\,°\text{C} = 273\,\text{K})$$

and pressure

$$(P^0_{\text{gas}} = 1 \text{ atmosphere} = 1.013 \times 10^6 \text{ dynes cm}^{-2}).$$

We know that the corresponding volume V^0 is 22.4 litres (22400 cm^3), and that the total number of hydrogen atoms contained in this volume is equal to Avogadro's number, $N_{\text{A}} = 1/m_{\text{H}} = 6.02 \times 10^{23}$. Under these conditions, the number density n is equal to $n_0 = N_{\text{A}}/V^0 = 2.69 \times 10^{19}$ cm^{-3}, and we find

$$k = P^0_{\text{gas}}/n_0 T^0 = 1.38 \times 10^{-16} \text{ c.g.s.} \tag{32}$$

Equation (31) then gives, in c.g.s. units,

$$R = kN_{\text{A}} = 8.32 \times 10^7 = \text{'gas constant'}. \tag{33}$$

If, instead of considering Equation (26) or Equation (30), which are valid for any mass of gas, we consider the more particular case of a mass equal to N_{A} particles of the mixture (one mole), this mass will be $(\mu m_{\text{H}})N_{\text{A}} = \mu$ grams; and if we suppose that it occupies a volume V, we will have $n = N_{\text{A}}/V$, so that the expression (26) for P_{gas} will take on the classical form of the perfect gas law for *one mole*:

$$P_{\text{gas}} V = RT. \tag{34}$$

In stellar interiors the atoms are almost completely *ionized* and we can assume, in the first approximation, that we are dealing with a state of *complete* ionization.

Let us examine the effect of such an ionization on the number density of free electrons and on the value of μ. Following the usual notation, we shall let X be the mass in grams of hydrogen contained (before ionization) in 1 gram of the mixture, and Y be the mass in grams of helium contained (before ionization) in 1 gram of the mixture. Finally, let $Z = 1 - X - Y$ be the mass of all the elements 'other than H and He' contained in 1 gram of the mixture. We shall say, somewhat loosely, that the sum of these 'other elements' represents the 'heavy elements'. We can then draw up the following table, in which lines (1), (2), (3), etc., give respectively:

(1) the mass in grams of H, He, and 'heavy elements' in one gram of the mixture;
(2) the mass in grams of *one* nucleus of a certain 'species';

(3) the number of nuclei of this species in one gram of the mixture;
(4) the number of electrons liberated by each completely ionized *atom*;
(5) the number of electrons liberated by the atoms of each species present in one gram of the mixture;
(6) the total number of electrons and nuclei for each species;
(7) the total number of particles of all kinds per gram of the mixture;
(8) the total number of free electrons per gram of the completely ionized mixture.

	Hydrogen	Helium	'Heavy elements'	Notes
(1)	X	Y	Z	
(2)	$m_H = 1/N_A$	$4m_H$	$\mu_Z m_H$	a
(3)	$X/m_H = XN_A$	$\frac{1}{4}YN_A$	$(1/\mu_Z)\, ZN_A$	
(4)	1	2	$\frac{1}{2}\mu_Z$	b
(5)	XN_A	$\frac{1}{2}YN_A$	$\frac{1}{2}ZN_A$	
(6)	$2XN_A$	$\frac{3}{4}YN_A$	$\frac{1}{2}ZN_A$	c
(7)		approximately $(2X+\frac{3}{4}Y)\, N_A$		d
(8)		$\frac{1}{2}(2X+Y+Z)\, N_A = \frac{1}{2}(1+X)\, N_A$		e

[a] Here we denote by μ_Z the mean mass of the 'heavy elements' expressed in units of m_H, and we neglect the mass of the electrons belonging to an atom in comparison with the mass of the nucleus.
[b] The nucleus of a neutral atom contains as many protons as there are orbital electrons. But in the nucleus there are *approximately* as many neutrons as protons. If the mean atomic mass in units of m_H is μ_Z, this means that the sum of the number of protons and the number of neutrons is equal to μ_Z. Therefore there are approximately $\mu_Z/2$ protons in the nucleus and $\mu_Z/2$ orbital electrons.
[c] When μ_Z is greater than or equal to 16, the term ZN_A/μ_Z will be negligible in comparison with the remaining term, equal to $\frac{1}{2}ZN_A$.
[d] Since X is of the order of 0.74, Y of the order of 0.24, and Z of the order of 0.02, we can neglect $Z/2$ in comparison with $(2X+\frac{3}{4}Y)$.
[e] Here we use the fact that $(X+Y+Z)=1$ by definition.

According to the result given in the next to last line of the table, the mean mass (in grams) $\bar{m}_g$ of a particle of the mixture is given by:

$$\bar{m}_g = (1 \text{ gram})/(\text{number of particles}) = 1/N_A\,(2X + \tfrac{3}{4}Y). \tag{35}$$

Consequently, the definition of μ given in (28) yields (taking into account the fact that $m_H N_A = 1$):

$$\mu = \bar{m}_g/m_H = 1/(2X + \tfrac{3}{4}Y). \tag{36}$$

Considering Z to be negligible in comparison with Y, we can replace Y by $(1-X)$, giving as a final result

$$\mu = (\tfrac{3}{4} + \tfrac{5}{4}X)^{-1}. \tag{37}$$

Moreover, since each cm^3 of the mixture contains by definition ϱ grams, the number N_e of free electrons per cm^3 is obtained by multiplying the last line of the table by ϱ, whence

$$\boxed{N_e = \varrho\,(1 + X)\, N_A/2}. \tag{38}$$

Remarks

(1) The perfect gas law is applied to stellar interiors because the high temperature of the densest regions causes them to be very far from the 'critical conditions' for liquefaction of the gas, despite the enormous pressures that prevail.

(2) The gas pressure P_{gas} is only one part (although generally strongly predominant) of the total pressure P, which results from the combined effect of gas pressure (bombardment by material particles) and of radiation pressure (bombardment by photons). The calculation of P_{rad}, taking into account the properties of the *radiation field* in the interior of a star (cf. [9], Chapter VII, Section 6.6), shows that to a very good approximation we have

$$\boxed{P_{rad} = \tfrac{1}{3}aT^4}, \tag{39}$$

where a is a constant equal to 7.56×10^{-15} c.g.s.

When we have calculated P_{gas} and T, Equation (39) will show that P_{rad} is very small in comparison with P_{gas} in the interior of a star like the Sun. In fact, we shall find that although the ratio P_{rad}/P_{gas} increases towards the center of the Sun, it never exceeds 6×10^{-4}.

6. A Model of the Sun at 'Constant Density' $\varrho = \bar{\varrho}$

To get an idea of the order of magnitude of the principal parameters, let us try to construct a 'model' of the Sun in the very first approximation (the only approximation that can be 'guessed') by assuming that the density of the Sun is independent of the distance r from the center, and therefore equal to the mean density $\bar{\varrho}$ obtained by dividing the total mass of the Sun ($M = 2 \times 10^{33}$ g) by its volume ($V = \frac{4}{3}\pi R^3$), where $R = 7 \times 10^{10}$ cm is the solar radius. With these figures, the density $\bar{\varrho}$ is equal to 1.4 grams per cm^3. In this section we have $\varrho = \bar{\varrho} = \text{const.}$

Let us set $r' = r/R$. At the center $r' = 0$. At the surface $r' = 1$.

In this notation we have

$$M_r = \tfrac{4}{3}\pi r^3 \varrho = \tfrac{4}{3}\pi (R^3 r'^3)\, \varrho = M r'^3 . \tag{40}$$

Substituting this expression for M_r into the equation of mechanical equilibrium (18) and replacing r by $r'R$, we find

$$-\,\mathrm{d}P/\mathrm{d}r' = (GM/R)\, \varrho r' . \tag{41}$$

Let us denote by A the constant coefficient of r' on the right-hand side of Equation (41). Limiting ourselves to a single significant figure, we have

$$A = (7 \times 10^{-8}) \times (2 \times 10^{33}) \times (1.4) \times (7 \times 10^{10})^{-1} = 3 \times 10^{15} \text{ c.g.s.} \tag{42}$$

Let us integrate the equation

$$-\,\mathrm{d}P/\mathrm{d}r' = Ar' , \tag{43}$$

making use of the physically obvious condition $P(1)=0$ (zero pressure at the surface $r'=1$ of the Sun).

We find at once

$$P(r') = \tfrac{1}{2}A(1 - r'^2). \tag{44}$$

We immediately deduce that the pressure at the center ($r'=0$) of the Sun, P_c, must be of the order of

$$P_c = P(0) = \tfrac{1}{2}A = 1.5 \times 10^{15}\ \text{dynes/cm}^2 = 1.5 \times 10^9\ \text{atm}. \tag{45}$$

Thus the pressure at the center appears to be of the order of 1.5 billion atmospheres. Now, a more exact calculation which we shall make later on gives 220 *billion atmospheres*. The error obviously results from the fact that $\varrho(r')$ actually varies with r', and increases *slowly* towards the center. In reality, we shall later find that $\bar{\varrho}=\varrho(0.45)$. Nevertheless, we note that the relative error in (45) is fairly small.

It is useful to put Equation (44) in the form

$$P(r') = P_c(1 - r'^2), \tag{46}$$

which shows explicitly that the pressure *decreases* from P_c to zero when one goes from the center C to the surface.

Despite the rough nature of the approximation $\varrho=\bar{\varrho}$, this hypothesis not only provides a reasonable order of magnitude for P_c, but also makes it possible (with the addition of a few plausible hypotheses concerning the composition of the stellar mixture and its physical state) to determine the order of magnitude of the central temperature T_c and the temperature distribution $T(r')$.

To begin with, let us make the additional hypothesis that the Sun is *composed of pure hydrogen* entirely dissociated into protons and electrons (*complete ionization*).

In this case, the 'mean mass μ of a particle' of the 'mixture' in units of m_{H} equals, according to (36) (or by 'direct' reasoning),

$$\mu = \tfrac{1}{2} \quad (\text{since } X = 1 \quad \text{and} \quad Y = 0); \tag{47}$$

and according to (29) the number density n in particles per cm^3 equals

$$n = \frac{\varrho}{\mu m_{\text{H}}} = \frac{1.4}{0.5 \times (6 \times 10^{23})^{-1}} = 2 \times 10^{24}\ \text{particles cm}^{-3}. \tag{48}$$

Neglecting radiation pressure and using relation (26) between P and T, we immediately find from the approximate formula (46)

$$T(r') = \frac{P(r')}{nk} = \frac{P(r')}{(2 \times 10^{24}) \times (1.4 \times 10^{-16})} = T_c(1 - r'^2), \tag{49}$$

with

$$T_c = T(0) = P_c/nk = 5 \times 10^6\ \text{K}. \tag{50}$$

This value of T_c is not very different from the value given by the most exact calculation (14.6×10^6 K). The order of magnitude is entirely correct. This in itself justifies the use of the perfect gas law (26) ('agitation' balancing compression), and to a certain extent it justifies and explains the validity of the hypothesis $\bar{\varrho} = \varrho$ (the center is more 'compressed' but hotter). However, this mention of 'compression' – very frequent in books devoted to internal structure – is worth a brief commentary.

Indeed, the mechanical equilibrium of a star is often presented as the result of a balance between the 'compression' of each layer ($\mathrm{d}V$) by the layers located above ($\mathrm{d}V$) and the *gradient* of the gas pressure (when one does not make the additional error of speaking of the pressure without qualification).

Now, the gradient of P does have a 'supporting' effect, but what it balances is not the supposed 'compression' (for according to Newton's theorem the effect of the outer layers is simply *zero*), but the attraction exerted on ($\mathrm{d}V$) by M_r, which *pulls* ($\mathrm{d}V$) towards the center (the gravitational force on $\mathrm{d}V$). Let us recall once more the physical reason for which the effect of the outer layers is zero: when the star is divided (Figure 5) into two parts (A) and (B) by a plane tangent to the layer ($\mathrm{d}V$) (where A is the part not including the 'nucleus' of mass M_r), part (A), which is close to ($\mathrm{d}V$), has a mass smaller than the 'outer' part of (B), but the latter compensates for this (exactly, according to Newton) by being located farther away on the average. Now (A) 'pulls' towards the outside, while (B) 'pulls' towards the center.

It would be easy to obtain a more realistic result by replacing the hypothesis that the Sun is composed of pure hydrogen ($X=1$, $Y=0$) with an hypothesis taking into account the observed values (at the surface) of X and Y. With $X=0.75$ and $Y=0.25$, we find (assuming $\varrho = \bar{\varrho}$) $T_c = 7.5 \times 10^6$, a slightly more accurate value than that given by (50).

As we see, the model constructed by assuming that the density is independent of the distance r' from the center provides an entirely correct idea of what takes place in the interior of the Sun, revealing the enormous magnitude of the central temperature (which was not at all obvious *a priori*). But this model does not explain the origin of this enormous temperature. Nor does it explain the origin of the energy ($L = 4 \times 10^{33}$ ergs) which the Sun emits each second, as it has been doing for at least 5 billion years. Now, these two observations suggest that the central region of the Sun behaves like a 'furnace' where there is 'burned' a fuel both very 'energetic' and almost inexhaustible.

The missing explanation will of course be provided by thermonuclear reactions. But these reactions, in turn, are possible only because of the very high temperature prevailing near the center. Here again we should warn the beginning reader against a very common error. It is not sufficient for the temperature to be very high, in order for thermonuclear reactions to take place; the temperature must be 'optimum'.

Indeed, the protons from the ionized hydrogen repel each other electrostatically (charges of the same sign). Only a sufficient degree of thermal agitation (high temperature) can give them a high enough velocity to overcome this repulsion, which increases as $1/d^2$ when the distance d between the protons decreases.

In fact, the repulsion increases indefinitely when d tends to zero, but at very small distances the forces of attraction (nuclear forces) come into play. The problem is to attain a distance d_p sufficiently small to enter the domain of the nuclear forces (whose range is very short); therefore T must be high enough so that the number of protons having velocities v large enough to penetrate to the distance d_p will not be negligible. But at the same time, the 'probability of fusion' of the protons which have come together is greater for those having *small* relative velocities (protons which remain 'in the same neighbourhood' for a longer time), and which are thus not *too* greatly 'agitated'. Hence the existence of the 'optimum temperature' to which we have just referred: it must be neither too low nor too high. It happens that for the Sun this optimum temperature is about 10×10^6 K. We shall study this question in more detail, and we shall give physically more rigourous explanations in Chapter IV.

7. The 'Homologous' Model. Expressions for P_c and T_c in Terms of M and R

Instead of making the hypothesis that ϱ is independent of r' (a homogeneous star), we can make another simplifying hypothesis which is less restrictive and which provides two very interesting immediate results concerning the relation of the central pressure P_c and the central temperature T_c to the total mass M and the radius R of the star.

Let us suppose that the density distribution, in relative value, is the same in all stars. In other words, let us suppose that if $\bar{\varrho}$ is the mean density, equal to M divided by the volume V of the star, we have (again setting $r' = r/R$):

$$\varrho(r')/\bar{\varrho} = D(r') \quad \text{or} \quad \varrho(r') = \bar{\varrho} D(r'), \tag{51}$$

with $D(r')$ (the density distribution function) the same for all stars (a 'universal' function). The universality of $D(r')$ establishes a sort of 'homology' among the set of all stars.

We immediately deduce, according to (19) (with ξ' denoting the running value of r'), that

$$M_r = \int_0^{r'} \bar{\varrho} D(\xi') \cdot 4\pi (R\xi')^2 R \, \mathrm{d}\xi' . \tag{52}$$

But since

$$\bar{\varrho} 4\pi R^3 = 3M , \tag{53}$$

it follows that if we denote by $f_M(r')$ the integral

$$f_M(r') = \int_0^{r'} 3D(\xi') \, \xi'^2 \, \mathrm{d}\xi' , \tag{54}$$

we have

$$M_r = M f_M(r'), \tag{55}$$

where $f_M(r')$ is a new 'universal' function valid for all stars, just like $D(r')$. Applying

(55) to the Sun, denoting by $M_\odot$ and $R_\odot$ the mass and radius of the Sun, and introducing the notation $M'(r') = M_r/M$, we have

$$M'(r') = f_M(r') = M'_\odot(r'). \tag{56}$$

This result is physically obvious. If the density is distributed in the same way in all stars, the ratio of the mass M_r to the total mass (which we have called M') will be distributed in the same way in all stars, and this distribution will also be valid for the Sun.

On the other hand, the two results which we shall now establish are not in any way 'intuitive'.

According to (18), in which we express M_r and ϱ in terms of the 'universal' quantities $M'(r')$ and $D(r')$, we have for all the 'homologous' models

$$\mathrm{d}P = -\frac{GM'(r')\,M}{R^2 r'^2}\,\bar{\varrho}D(r')\,R\,\mathrm{d}r'\,; \tag{57}$$

that is, using the expression for $\bar{\varrho}$ in terms of M and R,

$$\mathrm{d}P = -\frac{3}{4}\frac{G}{\pi}\frac{M^2}{R^4}\,M'(r')\,D(r')\,\frac{\mathrm{d}r'}{r'^2}. \tag{58}$$

Integrating, and making use of the boundary condition $P(1) = 0$, we immediately find

$$P(r') = \frac{M^2}{R^4}\,f_P(r'), \tag{59}$$

where $f_P(r')$ is a new 'universal function' defined by

$$f_P(r') = \frac{3}{4}\frac{G}{\pi}\int_{r'}^{1} M'(\xi')\,D(\xi')\,\frac{\mathrm{d}\xi'}{\xi'^2}. \tag{60}$$

If we apply Equation (59) to the Sun and divide (59) by the equation thus obtained, and if we let $P_\odot$ denote the pressure prevailing in the Sun at the depth r', it follows that

$$P(r')/P_\odot(r') = (M/M_\odot)^2/(R/R_\odot)^4. \tag{61}$$

In particular, applying (61) to the center $(r' = 0)$ and letting P_c and $P_{c,\odot}$ be the central pressure in a star and in the Sun, respectively, we have

$$P_c = P_{c,\odot}\,\frac{(M/M_\odot)^2}{(R/R_\odot)^4}. \tag{62}$$

The calculation of the temperature distribution (neglecting radiation pressure) introduces the number density n through Equation (26) – that is, according to (29) it introduces the parameter μ, which depends only on X and Y (since it is given by (36)). Thus, by adding to all the preceding hypotheses ('homology') the hypothesis that X and Y are *independent* of r' (corresponding to a 'stage of evolution' which has not 'advanced' very far from the moment at which the 'transmutation' of H into He begins *near the center*), we find, by an argument analogous to that made for P_c,

$$T_c = T_{c,\odot} \frac{(M/M_\odot)}{(R/R_\odot)}. \tag{63}$$

More generally, we have

$$T(r') = f_T(r')\,\mu(M/R), \tag{64}$$

where $f_T(r')$ is a certain 'universal' function of r'.

Exercise 1

N.B. – In the following problem we are interested only in *orders of magnitude*. Consequently, all the calculations can be made to a *maximum* of *two* significant figures (using a slide rule).

At some initial time t_0, we consider a spherical gaseous mass (S) of neutral hydrogen, having a mass of 10^6 solar masses and a volume of 1 pc^3 (3×10^{55} cm^3). In the first approximation, we assume that this mass is homogeneous and has a uniform temperature of 10^4K.

(1) Calculate the radiative energy density u_{rad} and the kinetic energy density (of thermal motion) u_{kin} in (S) at the time t_0.

Is it permissible to consider (S) as being in a state of *thermodynamic equilibrium* at the time t_0? (We recall that $a \approx 10^{-14}$ c.g.s.)

(2) Show that at the time t_0 the sphere (S) cannot be in *mechanical* equilibrium, and that it must collapse under the effect of the gravitational attraction of its particles. Find (assuming that S remains homogeneous and isothermal) the duration τ of its collapse to a negligibly small final radius. Express τ in years. Compare τ with the order of magnitude of the age of the universe. Indicate – without calculations, but by a qualitative and purely physical argument – whether abandoning the hypothesis of homogeneity and replacing it with a physically more plausible hypothesis would have the effect of increasing or decreasing τ.

(3) We assume that initially all the kinetic energy produced by compression goes into the (collisional) ionization of the hydrogen, and that the gas remains at 10^4K. We recall that the ionization potential of hydrogen is 13.5 eV.

(3a) Find the energy consumed by the ionization of one gram of neutral hydrogen, making use of the well-known ionization potential of hydrogen.

(3b) Find the energy produced by the *isothermal* compression of one *gram* of neutral hydrogen, in terms of the density ϱ_0 at the time t_0 and of the density ϱ_1 at the end of the isothermal compression phase.

(3c) We assume that the compression ceases to be isothermal at the time t_1 at which the energies calculated in (a) and (b) become equal. Find the corresponding value of ϱ_1, and show that this is precisely the density at which u_{rad} becomes of the same order as u_{kin}.

(4) At the time t_0, find the order of magnitude of the number N_2 of atoms of neutral hydrogen per cm^3 in the quantum state $n=2$, using the information that this state is located about 10 eV above the ground level. (*N.B.* – We shall not take into account the difference in statistical weight between these two levels. We recall that $k \approx 1.4 \times 10^{-16}$ c.g.s.) Can the usual formulas be employed under the physical conditions found in question (1)?

(5) Explain why the photons φ_1 corresponding to the transition $n=2 \to n=1$ have no chance of escaping from (S) in a reasonable time.

(6) Determine the mean free path L_2 of the photons φ_2 arising from the transition between the ionized state and the neutral state $n=2$, at the time t_0, using the fact that the microscopic ionization cross-section from the level $n=2$ is of the order of 10^{-17} cm^2.

Exercise 1′

(1) Review the proof of Poisson's equation for a star (S) formed of spherical homogeneous layers, assuming that Newton's theorem concerning the gravitational field inside (S) is known, and making the argument for a *complete* layer (r, dr) whose center is the center O of (S).

(2) Show that Poisson's equation can also very easily be established by making the argument for the *truncated conical* element cut out of this layer by the cone whose center is O and whose aperture is the solid angle $(\mathrm{d}\omega)$.

(3) What are the complications encountered in the proof when the argument is made for a *cylindrical* element whose generatrix is parallel to one of the radii of (S), instead of for a truncated conical element?

(4) Show that one can nevertheless recover the classical form of Poisson's equation, although with greater difficulty, by making the argument for a *cylindrical* element – taking care not to neglect any important first-order term.

(5) Similarly, show that in spite of some very instructive complications, one can also recover the relation between the gradient of the total pressure and the gravitational force per unit volume in a completely rigourous fashion by making the argument not for a *cylindrical* element but for the truncated conical element already considered in question (2).

Exercise 2

Inspired by an article in the *Journal of Geophysical Research* **70** (1965) 3819–29.

The problem below is related to the question raised when one examines the possibility of a volcanic origin for the dust particles that cover certain parts of the Moon. The study is based on recent progress in our knowledge of the phenomenon of the 'fluidization' of solid particles (dust particles, grains of sand, etc.) by a gas flow. In the 'dense' phase of fluidization by a vertical gas flow, the solid particles are *maintained* at a fixed altitude, despite their weight, by the upward motion of the gas. At the beginning of the problem we shall assume that this 'dense phase' is present. Throughout the problem we shall assume that the gaseous agent is water vapour, chemically pure (except for the presence of solid particles) and at a constant temperature $T = 1120\,\mathrm{K}$ (constant in time and space). We recall that the gas constant is $R = 8.32 \times 10^7$ c.g.s., and that the mean value of the acceleration of gravity is 980 c.g.s. at the surface of the Earth and 162 at the surface of the Moon (these quantities will be assumed to be constant throughout the problem). The mean density of each individual grain, ϱ_{00}, will be assumed to be 2.4 grams per cm^3.

(1) We consider the flow of volcanic ash, fluidized by steam, in a cylindrical chimney located near the surface of the Moon or of the Earth. The chimney is narrow enough so that its dimensions other than vertical length can be neglected, but wide enough so that the friction of the mixture against the walls can also be neglected. We denote by y the height in cm of a point in the mixture above the point O, the origin of 'height'. At the point O the prevailing pressure is P_0 (we shall call this pressure $P_{0,E}$ or $P_{0,M}$ when we want to refer specifically to the Earth or to the Moon). Let ε be the fraction of the volume of the mixture occupied by the gas. For the 'dense' phase $\varepsilon < 0.8$. This quantity ε is a function $\varepsilon(y)$ of y, zero for $y = 0$, which will be calculated (in the form of a numerical table) below (in paragraph 6).

(1.1) Establish a general formula (1), which expresses the mass ϱ_s of the (solid) dust particles per cm^3 of the mixture, in terms of ϱ_{00} and of ε.

(1.2) Let $P(y)$ be the function describing the variations of the *gas* pressure as a function of y. Write Equation (2), which expresses the equilibrium of the *dust particles* under the combined influence of their weight and of the gas flow. (Your equation will not be rigourous, but it will be an excellent approximation.)

(2) We set $y' = (g\varrho_{00}/P_0)\, y$, and $P' = P/P_0$. Derive from Equations (1) and (2) a very simple formula (2′), which will make it easy to compute a table of values of y' as a function of ε once we have a table of the function $P'(\varepsilon)$. Specify the procedure to be followed in such a computation.

(3) We assume that the total *mass* of the mixture contained in a chimney above each cm^2 of the level $y = 0$ is the same for the Moon and for the Earth. Calculate $P_{0,M}$ in terms of $P_{0,E}$, the atmospheric pressure $P_{A,E}$, and some of the numerical data given above.

(4) Let ϱ_g be the mass in grams of the gas contained in 1 cm^3 *of the gaseous component* of the mixture, and $\varrho_g(P')$ the mathematical function giving the density of the gas in terms of the value of P' corresponding to a certain level y. Similarly, let $v(P')$ be the velocity v of the gas at the level y, as a function of P'. We assume that each gram of *dust* emits, by a sort of 'outgassing', q grams of steam per second (in the numerical applications we shall take $q = 3.55 \times 10^{-7}$ c.g.s.)

(4.1) Express by an Equation (3) the fact that the situation remains stationary at the level y, insofar as the mass of *gas* (steam) to be found between y and $(y + \mathrm{d}y)$ is concerned.

(4.2) We assume that $v(1)=0$. Find εv in terms of P' and of the parameters g, q, T, R, and μ (where μ is the molecular mass of the gas in units of m_H).

(4.3) Show that for the Earth $v(P')$ is given by the approximate relation:

$$\varepsilon v = (15/8)\,[(1/P') - 1]\,. \tag{4}$$

(4.4) Write the corresponding formula (5) for the Moon.

(5) We assume that the flow of the steam through the porous medium formed by the stationary 'solid' component of the mixture creates a dynamical pressure gradient given by

$$\mathrm{d}P/\mathrm{d}y = -(v/8)(1-\varepsilon)^2/D^2\varepsilon^2, \tag{6}$$

where D is a mean diameter of the dust particles, which are assumed to be spherical (in the numerical applications, $D=0.01$ cm).

(5.1) Find P' as a function of ε and of the parameters g, q, ϱ_{00}, μ, D, R, and T.

(5.2) Put this function in the form of a fraction with $(1-\varepsilon)$ as the numerator.

(5.3) Find an expression for the constant F which then appears in the denominator, whose form is $[(1-\varepsilon)+F\varepsilon^n]$, for the Earth and for the Moon. (In all of question (5), and below, we assume that $v(1)=0$.)

(6) The numerical values of the function $P'(\varepsilon)$, for the Earth and for the Moon, are given in the following table:

ε	0.00	0.10	0.20	0.30	0.40	0.50	0.60	0.64	0.68	0.72	0.76
$1000\,P'_E$	1000	999	990	962	904	799	649	578	504	429	353
$1000\,P'_M$	1000	1000	1000	999	997	993	985	980	974	965	952

$P'_E(0.80)=0.281$ and $P'_M(0.80)=0.934$.

(6.1) Use the method found in question (2) to draw up a table of P' as a function of y' for the Earth and for the Moon, corresponding to the values of ε in the above table.

(6.2) Find the value of $P_{0,E}$, assuming that $\varepsilon=0.80$ for the Earth corresponds to the upper boundary of the mixture which is subject to a pressure of 10^6 c.g.s. in the terrestrial atmosphere.

(6.3) Let $y(\varepsilon)$ be the inverse function of $\varepsilon(y)$. Find $y(0.8)$ for the Earth and for the Moon.

(6.4) What is the value of $P(0.8)$ for the Moon, where $P(\varepsilon)$ represents P as a function of ε?

(7) When ε exceeds 0.80, we pass from the 'dense' phase to the 'dilute' phase, in which the dust particles no longer remain stationary but are carried along by the gas flow.

(7.1) How can the preceding calculation furnish a tentative justification for the hypothesis of volcanic origin of the dust on the surface of the Moon?

(7.2) What is the alternative hypothesis?

Exercise 2′

We consider a cylindrical enclosure (A) with a *horizontal* base and a vertical generatrix, whose cross-section is a few meters and whose altitude is of the order of 5 meters. The enclosure (A) is hermetically sealed and protected from any exchange of heat with the outside. It is located in France, at sea level ($g=981$ c.g.s.). We denote by z the altitude above sea level in cm.

We also consider a gaseous sphere (S) of radius $R=7\times10^{10}$ cm and of mass $M=2\times10^{33}$ grams, which is too cold to be the site of thermonuclear reactions. To avoid confusion with R, we shall call the gas constant (k/H). We recall that its value is 8.32×10^7 c.g.s. The constant G is equal to 6.67×10^{-8} c.g.s. We denote by $\varrho(r)$ the density at a distance r from the center, and by $T(r)$ the corresponding temperature.

(1) Recall the numerical values of the exponents a and b – which, in the first approximation, are assumed to be *integers* – in the formula

$$\bar{k} = k_0\,\varrho^a T^{(b+1/2)},$$

which gives the opacity in cm^2 per cm^3. We shall retain these values of a and b throughout the following problem (on this subject, see [9], Chapter VII, Equations (51) and (113)).

(2) Explain very briefly the physical mechanism by which an incompressible, isothermal ($\mathrm{d}T/\mathrm{d}z=0$) liquid at rest, which is assumed to be *initially* homogeneous ($\mathrm{d}\varrho/\mathrm{d}z=0$) and which completely or partially fills (A), *can* remain in mechanical equilibrium from the outset.

(3) Explain why the same 'initial conditions' ($\mathrm{d}T/\mathrm{d}z=0$, $\mathrm{d}\varrho/\mathrm{d}z=0$) applied to a perfect gas (e.g., molecular nitrogen, $\mu=28$) which completely fills the enclosure (A) do not enable the gas to remain in equilibrium from the outset.

(4) The nitrogen contained in (A), which is assumed to be initially isothermal and homogeneous, is therefore necessarily set in motion before it attains its definitive state of mechanical equilibrium. Without formulating any equations, describe the initial phases of this motion for the various layers, neglecting viscosity and any heating of the gas. Show by as clear a physical argument as possible that the situation must converge towards a state of stable equilibrium characterized by final distributions $\varrho(z)$ and $P(z)$ such that $\mathrm{d}\varrho/\mathrm{d}z<0$ and $\mathrm{d}P/\mathrm{d}z<0$. ($P$ denotes the gas pressure; the radiation pressure will always be neglected.)

(5) Calculate $\varrho(z)$ and $P(z)$ in terms of the physical parameters which are applicable to the case of nitrogen. Give (without skipping over the details of the calculation) the final answer in the form of formulae which contain only z and a minimum number of numerical coefficients. We assume that $T=0°\mathrm{C}$.

(6) We assume that (A) is filled with dry nitrogen at 0°C in such a quantity that $P(0)=10^6$ c.g.s. (at the level $z=0$), and we introduce into (A) at the level $z=0$ a little plastic balloon, whose volume is about one liter and which contains molecular hydrogen ($\mu=2$). We assume that this action does not disturb the initial conditions of this new experiment, the distribution $\varrho(z)$ and $P(z)$ calculated above. Finally, we assume that the balloon can expand without breaking throughout the experiment, and that in expanding it creates only negligible elastic forces in its envelope, whose mass is assumed to be negligible. (*N.B.* – The balloon is at 0°C.)

Describe quantitatively the principal phases of what happens when the initial density of the balloon is *three times* $\varrho(0)$ and when it is released in the enclosure (A) with no initial velocity. To what real experiments does this calculation correspond? What practical precautions does it suggest?

(7) We now return to the sphere (S), and in particular we consider a sphere $(S)_\varrho$ such that initially, at rest, we have $\mathrm{d}\varrho/\mathrm{d}r\neq 0$ and $\mathrm{d}T/\mathrm{d}r=0$. In other words, the gas is isothermal but not homogeneous. Can a sphere $(S)_\varrho$ be in mechanical equilibrium from the outset?

(8) Next we consider a sphere $(S)_T$ such that initially, at rest, we have $\mathrm{d}\varrho/\mathrm{d}r=0$ (a homogeneous sphere) but $\mathrm{d}T/\mathrm{d}r\neq 0$ (not isothermal).

(a) In this case, what must be the constant value ϱ_0 of ϱ? *Prove* the answer (which is easily guessed) by using the classical equations.

(b) Show that in this case, the distribution of mass and pressure as a function of $r'=r/R$ is completely determined, and depends only on parameters already given above. Find the functions

$$M'=M_r/M \quad \text{and} \quad P(r').$$

What is the *numerical* value of the central pressure P_c? (We assume that the sphere contains only hydrogen and helium in the proportions $X=0.75$, $Y=0.25$ grams per gram.)

(c) Determine the temperature distribution $T(r')$ and the central temperature T_c (a *numerical* value!). Explain why we find a temperature of the same order at the center of the Sun. Does this surprise you? Is it a case of compensating errors? What hypothesis is contradicted (to a certain extent) by the value of T_c?

(d) Show that the hypothesis $\varrho=\varrho_0$ violates one of the fundamental laws which the gas in the sphere $(S)_T$ should obey, considering the hypotheses made above. Specify numerically the 'non-constancy' of a certain parameter which should remain constant, by calculating it for $r'=0$, $r'=\frac{1}{2}$, $r'=\frac{3}{4}$, and $r'=1$.

(9) We consider a sphere (S) which is initially isothermal and homogeneous. Show that it cannot remain in equilibrium and explain why it cannot spontaneously become either $(S)_\varrho$ or $(S)_T$, but must evolve towards a structure like that of the Sun. Can the parameter ε be non-zero without nuclear reactions? Can the radius R be conserved in the new structure?

CHAPTER III

THE DETERMINATION OF THE INTERNAL STRUCTURE BY THE DENSITY DISTRIBUTION $\varrho(r)$

1. Introduction

In the light of the relations established in the preceding chapter, it begins to be clear that the study of the internal structure of a star like the Sun consists in discovering how certain physical quantities like the density ϱ, the temperature T, the total pressure P, etc., vary as functions of the distance r from the center.

Mathematically, this is equivalent to seeking a set of unknown functions $\varrho(r)$, $T(r)$, $P(r)$, etc., such that all the physical relations involving these quantities are satisfied. Most of these physical relations are represented by differential equations. We must therefore solve a system of differential equations, taking into account a certain number of boundary conditions, of which some are 'theoretical' (for example, $P(R)=0$), while others are imposed by the observations (for the Sun, $M_r(R)=M_\odot = 1.99 \times 10^{33}$ grams and $R=R_\odot=6.96 \times 10^{10}$ cm).

In particular, we shall have to use the values of the 'abundances' X, Y, and Z of hydrogen, helium, and 'heavy elements' at the surface of the Sun $(r=R)$ which are provided by the quantitative analysis of spectra that refer essentially to the 'solar atmosphere'. Let us indicate these values right away: $X(R)=0.744$, $Y(R)=0.236$, $Z(R)=0.020$; also, the 'luminosity' $L(R)=3.90 \times 10^{33}$ ergs/sec.

The integration of this set of differential equations is not very difficult when modern electronic machines are available. But finding a large *excess of physical parameters* capable of describing the same state of the Sun at a distance r from the center, we are faced with two preliminary problems:

(a) the choice of the 'fundamental' unknown functions – knowing which we can entirely determine the distribution of all the other parameters;

(b) a search for the 'physical facts' which set the conditions for the existence and the uniqueness of the solution.

Concerning the first of these problems, we shall see in the present chapter that the function $\varrho(r)$ is precisely one of these fundamental functions. Knowing it, we can determine the variation of M_r, P, and T with r in such a way that the following equations are all satisfied, as well as the boundary conditions, at least in the region in which r varies from $0.3R$ to R (for a star like the Sun):

$$\mathrm{d}M_r/\mathrm{d}r = 4\pi r^2 \varrho\,, \tag{1}$$

$$\mathrm{d}P/\mathrm{d}r = -\,GM_r r^{-2} \varrho\,, \tag{2}$$

$$P = P_{\mathrm{gas}} + P_{\mathrm{rad}}\,, \tag{3}$$

$$P_{\text{gas}} = \frac{k}{m_{\text{H}}} \frac{\varrho}{\mu} T\,, \tag{4}$$

$$P_{\text{rad}} = \tfrac{1}{3} a T^4\,, \tag{5}$$

$$\mu^{-1} = 0.75 + 1.25X\,. \tag{6}$$

($k/m_{\text{H}}=8.32\times10^7$ c.g.s.; $a=7.56\times10^{-15}$ c.g.s.)

The restriction to the region $0.3R<r<R$ is imposed by the fact that the composition (X, Y, Z) can be modified (in a star where there is no 'mixing' between the center and the exterior) only by *nuclear reactions* which convert hydrogen to helium, decreasing X and increasing Y. But such reactions can take place (as we shall show in a chapter devoted to thermonuclear reactions) only when the temperature is high enough. Now, we shall find that the temperature of the Sun increases towards the center, but does not attain the value necessary for thermonuclear reactions until about $r=0.3R$. Thus throughout the region from $r=R$ to $r=0.3R$, we can adopt a constant value for X and Y independent of r, and we can take for X and Y the values 'observed' at the surface (0.744 and 0.236).

We will not be able to study the central region $(r<0.3R)$ until later, after we have become acquainted with the relation between the net output of radiation across a sphere Σ_r of radius r and the temperature gradient*, and established a certain number of properties of thermonuclear reactions (Chapter IV).

Naturally, we shall use the 'right' distribution $\varrho(r)$ to illustrate our analysis – the one resulting from the work of M. Schwarzschild (*The Structure and the Evolution of the Stars* [1]), which satisfies not only the six equations given above, but also all the other conditions of the problem which result from Chapter VII of [9] and from our Chapter IV below. In this way, the distributions we shall find for $P(r)$, $T(r)$, etc., will also be the 'real' pressure and temperature distributions in the Sun.

But we will not be able to solve problem (b) until the end of our Chapter IV, in the sense that only then will we have shown that our $\varrho(r)$ is really the 'right' distribution that satisfies *all* the conditions (concerning both mechanics and energy) of the problem.

Let us start, then, with the distribution described by Table I, and let us show how Equations (1) to (6) enable us to deduce from it the variations of M_r, P, and T.

2. The Determination of the Distribution of the Mass M_r Contained in a Sphere of Radius r

It is obvious that M_r can be calculated as a function of r by numerical integration of Equation (1), as soon as some distribution $\varrho(r)$ – like that in Table I – has been adopted.

But it is still of great interest, if only to have a better idea of the relative importance of the various contributions to the final result, to follow the details of such an integra-

* This relation is established in Chapter VII of *Introduction to the General Theory of Particle Transfer* by the present author, referred to as [9] in the following pages.

TABLE I

The density distribution in the Sun as a function of the distance from the center[a]

$r'=r/R$	0.0	0.1	0.2	0.3	0.4	0.5	0.6	0.7	0.8
ϱ (g/cm³)	135	86	36	13	4.1	1.3	0.40	0.12	0.036
$r'=r/R$	0.90	0.92	0.94	0.96	0.98	1.00			
$10^3\varrho$ g/cm³	9.45	6.55	4.14	2.16	0.752	0.00			

[a] Cf. [1], p. 259; watch out for the factor 1000 in the second part of the table.

tion, limiting ourselves to the accuracy that can be obtained by using a slide rule (to three significant figures) and a relatively large 'integration step', in order to decrease the number of operations in the computation. In giving the 'rough' result obtained in this way, we shall also note the more accurate result that would have been obtained by increasing the number of steps and replacing the slide rule with a desk calculator or a computer. Thus we shall avoid the accumulation of errors by using the 'exact' values in the following stage of the calculation, while maintaining a close contact with the significant details of the computation (order of magnitude of the intermediate factors).

Moreover, in numerical computations it is convenient to introduce 'dimensionless variables' whenever they have a sufficiently explicit physical meaning, which will be the case if we set

$$r' = r/R_\odot \tag{7}$$

and

$$M'_{r'} = M_r/M_\odot = M', \tag{8}$$

that is, if we measure distances and masses in units of the radius and the total mass of the Sun, respectively. Whenever there is no risk of confusion, we shall use the notation M', suppressing the subscript r'.

In terms of these new variables, Equation (1) can be written:

$$M_\odot \, \mathrm{d}M'/R_\odot \, \mathrm{d}r' = 4\pi r'^2 R_\odot^2 \varrho,$$

or

$$\mathrm{d}M'/\mathrm{d}r' = [4\pi R_\odot^3/M_\odot] \, \varrho r'^2. \tag{9'}$$

With the numerical values

$$R_\odot = 6.96 \times 10^{10} \quad \text{and} \quad M_\odot = 1.99 \times 10^{33} \text{ c.g.s.},$$

the constant in brackets equals 2.13 and the equation reduces to

$$\mathrm{d}M'/\mathrm{d}r' = 2.13 \, \varrho r'^2. \tag{9}$$

In order to simplify the numerical computation as much as possible, we shall use the 'trapezoid' method – that is, the working rule:

$$\begin{aligned} y(x+\Delta x) &= y(x) + \Delta y \\ \Delta y &= \tfrac{1}{2}\Delta x\,[y'(x) + y'(x+\Delta x)] \\ y'(x) &= \mathrm{d}y/\mathrm{d}x\,. \end{aligned} \tag{10}$$

Here we can begin the integration either at $r'=0$ or at $r'=1$; but as Table II shows, $\Delta M'$ varies less rapidly near $r'=1$. For this practical reason, it is better to begin the integration at $r'=1$, in order to accumulate fewer errors at the beginning. It is therefore appropriate to recompute Table II 'from top to bottom', beginning with the left-hand column and proceeding column by column. We strongly urge the reader *actually to redo* this calculation, using our table only as a check (and not too strict a check at that, since we ourselves have made this computation only with a slide rule, and a small numerical disagreement does not necessarily imply an error in the method).

TABLE II

Approximate calculation of the mass distribution $M'(r')$

$r'=r/R_\odot$	ϱ	$\mathrm{d}M'/\mathrm{d}r'$	$-\Delta M'$	M'_{app}	M'_{exact}
1.00	0.000	0.000		1.000	1.000
			0.001		
0.90	0.009	0.02		0.999	0.999
			0.003		
0.80	0.036	0.05		0.996	0.996
			0.009		
0.70	0.12	0.13		0.987	0.988
			0.022		
0.60	0.40	0.31		0.965	0.967
			0.050		
0.50	1.30	0.70		0.915	0.92
			0.105		
0.40	4.1	1.4		0.81	0.82
			0.19		
0.30	13	2.5		0.62	0.63
			0.28		
0.20	36	3.1		0.34	0.34
			0.25		
0.10	86	1.85		0.09	0.07
			0.09		
0.00	135	0.00		0.00	0.00

We note that:

(1) The method we have used gives M' with satisfactory accuracy, of the order of 0.02 in absolute value.

(2) Most of the total mass of the Sun is concentrated near the 'central half': M' remains very close to 1 from the surface to about $r'=0.60$.

(3) The increment $(-\Delta M')$ is of course always positive, but it passes through a maximum around $r'=0.20$ ($\Delta M'<0$ because $\Delta r'=-0.10$).

(4) The tangents at the origin ($r'=0$) and at the point $r'=1$ have zero slope. This

does not result from the calculation of $\Delta M'$, but from that of $\varrho r'^2$, which is proportional to $\mathrm{d}M'/\mathrm{d}r'$.

An even more precise and more elegant 'test' of the approximate numerical integration is indicated in Figure 10. It consists of showing that the curve representing the exact function $M'(r')$ does have tangents whose slope is given by the derivative $\mathrm{d}M'/\mathrm{d}r'$ computed according to (9).

3. The Determination of the Distribution of the Total Pressure *P* as a Function of *r*

Using the change of variables given by (7) and (8), we can put Equation (2) in the form

$$\mathrm{d}P/\mathrm{d}r' = -\left[GM_\odot/R_\odot\right] M' r'^{-2} \varrho \,, \tag{11}$$

which, with $G = 6.67 \times 10^{-8}$, $M_\odot = 1.99 \times 10^{33}$, and $R_\odot = 6.96 \times 10^{10}$ c.g.s., reduces to

$$\mathrm{d}P/\mathrm{d}r' = -\ 1.91 \times 10^{15}\ M' r'^{-2} \varrho \,. \tag{12}$$

Here, the integration can begin only at $r' = 1$ (the surface of the Sun), for we know the value of the pressure P only at this boundary, where it is obviously equal to *zero*. In the interval $1 > r' > 0.9$, the variation is so rapid that the integration step (Table III) must be reduced to $\Delta r' = -0.02$, if we want to obtain more or less correct values of P_{app}. Of course, we use here the exact values of M', in order to avoid the influence of accumulated errors.

Beginning with $r' = 0.90$, we can return to the step $\Delta r' = -0.10$, giving Table IV. As with the computation of M', an excellent 'test' is obtained by comparing graphically (Figure 11) the slopes $\tan\alpha$ of the tangents to the curve representing $\log_{10} P$, as given on the one hand by Equation (13)

$$\tan\alpha = \frac{\mathrm{d}\log_{10} P}{\mathrm{d}r'} = 0.434 \frac{\mathrm{d}\log_e P}{\mathrm{d}r'} = 0.434 \left(\frac{1}{P}\frac{\mathrm{d}P}{\mathrm{d}r'}\right) \tag{13}$$

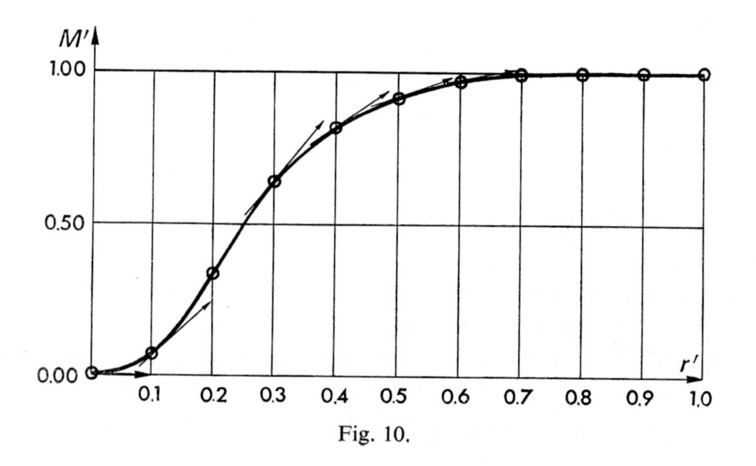

Fig. 10.

TABLE III

Approximate calculation of the pressure distribution

r'	M'	$-(\mathrm{d}P/\mathrm{d}r') \times 10^{-12}$	$\Delta P \times 10^{-10}$	P_{app}	P_{exact}
1.00	1.000	0.00		0	0
			+ 1.50		
0.98	1.000	1.50		1.5×10^{10}	1.2×10^{10}
			+ 5.97		
0.96	1.000	4.47		7.5×10^{10}	6.8×10^{10}
			+ 13.35		
0.94	0.999	8.88		2.1×10^{11}	2.0×10^{11}
			+ 23.48		
0.92	0.999	14.6		4.4×10^{11}	4.3×10^{11}
			+ 36.6		
0.90	0.999	22.0		8.1×10^{11}	7.9×10^{11}

TABLE IV

Approximate calculation of the pressure distribution

r'	M'_{ex}	$-10^{-15} \times \mathrm{d}P/\mathrm{d}r'$	$+\Delta P$	P_{app}	P_{exact}
0.90	0.999	0.02		8×10^{11}	8×10^{11}
			65×10^{11}		
0.80	0.996	0.11		73×10^{11}	62×10^{11}
			295×10^{11}		
0.70	0.988	0.48		37×10^{12}	31×10^{12}
			13×10^{13}		
0.60	0.97	2.07		17×10^{13}	14×10^{13}
			56×10^{13}		
0.50	0.92	9.10		73×10^{13}	61×10^{13}
			24.5×10^{14}		
0.40	0.82	40		32×10^{14}	27×10^{14}
			10.7×10^{15}		
0.30	0.63	174		14×10^{15}	12×10^{15}
			38×10^{15}		
0.20	0.337	590		52×10^{15}	46×10^{15}
			89×10^{15}		
0.10	0.073	1200		14×10^{16}	14×10^{16}
			60×10^{15}		
0.00	0.000	0		20×10^{16}	22×10^{16}

(where we use the approximate values of $\mathrm{d}P/\mathrm{d}r'$ and the exact values of P), and on the other hand (for intermediate values of r') as given by (13′):

$$\tan\alpha = \frac{\Delta \log_{10} P_{\text{exact}}}{|\Delta r'|} = +\, 10\Delta \log_{10} P_{\text{exact}} \,. \qquad (13')$$

Tables III, IV, and V and Figure 11 evoke the following observations:

(1) The relative errors of the approximate calculation are rather large, but the *order of magnitude* of the results obtained in this way is always correct; the error does not exceed 15%.

TABLE V

Values of $(+\tan\alpha)/10$ computed from the approximate values of $(\mathrm{d}P/\mathrm{d}r')$ and the exact values of P

$r'=$	0.9		0.8		0.7		0.6		0.5		0.4		0.3		0.2		0.1		0.0
$100(+\tan\alpha/10)_{\mathrm{app}}$	110		77		67		64		65		64		63		55		37		00
$100\Delta\log_{10}P_{\mathrm{exact}}$		90		70		65		65		64		64		60		47		21	

(2) The pressure at the center (or, as we say, the 'central pressure') P_c is enormous, for it reaches 220×10^{15} c.g.s. – that is, 220 *billion atmospheres* ($1\ \mathrm{atm}=1.013\times10^6$ c.g.s.).

(3) The pressure gradient very close to the surface (in the interval of r' from 0.98 to 1.00) is very large, since the pressure changes from zero to a value of 1.2×10^{10} c.g.s. in an interval $\Delta r'$ equal to only 0.02. Near the surface, we have $r'=1$ and $M'=1$, and the density ϱ is nearly zero. But the numerical factor (-1.91×10^{15}) is so great that a density of the order of 10^{-3} is sufficient to produce a variation ΔP of the order of 10^{10}, when r' varies only over $\Delta r'=0.02$. Nevertheless, this pressure of 1.2×10^{10} at $r'=0.98$ is only 6% of a millionth of the pressure at the center!

(4) As we see in Figure 11, the variation of $\log_{10}P$ in the range of r' from 0.25 to 0.75 (between $\frac{1}{4}$ and $\frac{3}{4}$ of the radius) is almost completely *linear*. In other words, in this region P varies as an exponential function of the form $e^{-\alpha r'}$, where α is of the order of 15 and is independent of r'.

(5) Replacing M' in the neighbourhood of $r'=0$ by its value $\frac{4}{3}\pi r'^3\varrho_c$ (where ϱ_c is the density at the center), we immediately find that the derivative $\mathrm{d}P/\mathrm{d}r'$, which is proportional to $M'\varrho_c/r'^2$, can be represented near the center by const $r'\varrho_c^2$, and consequently tends to *zero*. Thus we see that according to (13) (with $P=P_c$ finite at the center), the tangent at the origin to the curve representing $\log_{10}P$ will be *horizontal*.

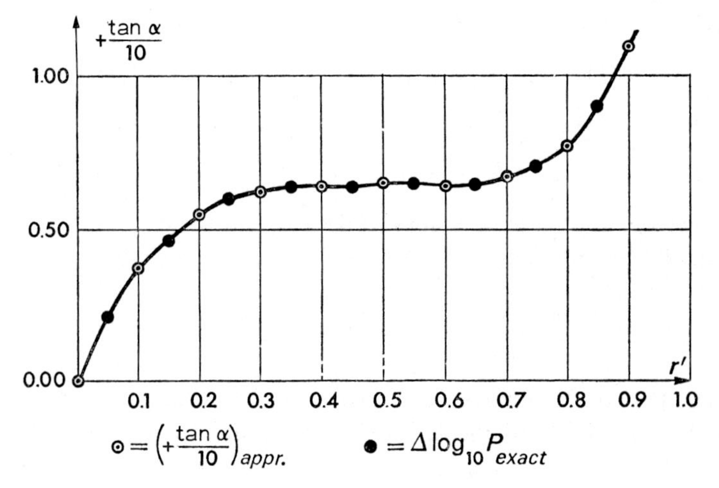

Fig. 11.

4. The Determination of the Distribution of the Temperature T as a Function of r

Equations (3), (4), (5) and (6), in which we consider X as a constant, make it possible to write the relation

$$P = P_{\text{gas}} + P_{\text{rad}} = (k/m_{\text{H}})\,(0.75 + 1.25X)\,\varrho T + \tfrac{1}{3}aT^4\,. \tag{14}$$

Since we already know P, ϱ, and $X = X_0 = 0.744$, this constitutes a fourth-order equation in T. But a rapid numerical analysis of this equation immediately reveals that because the coefficient a is so small (it equals 7.56×10^{-15}), the term in T^4 is negligible in the range of T in which we seek the solution of this equation. T can thus be determined by the linear equation

$$P = P_{\text{gas}} = (k/m_{\text{H}})\,(0.75 + 1.25X_0)\,\varrho T\,, \tag{15}$$

provided it is shown afterwards, as we shall do in a moment, that the corresponding values of $P_{\text{rad}}/P_{\text{gas}}$ are actually much less than unity.

When all the numerical coefficients are replaced by their c.g.s. values, Equation (15) reduces to

$$T = 7.2 \times 10^{-9}\, P/\varrho\,. \tag{16}$$

If, in order not to accumulate errors, we use the exact values of P and ϱ, this formula will give us an approximate expression for T (approximate in the sense that we have neglected P_{rad} in P, and that we have used all the way to the center a value of X equal to that observed at the surface of the Sun). Using Tables II, III, and IV, we obtain Table VI.

TABLE VI

Approximate calculation of the temperature distribution

r'	$\log_{10} P_{\text{ex}}$	$\log_{10} \varrho_{\text{ex}}$	T_{app}	T_{exact}
1.00	–	–	–	–
0.98	10.063	−3.123	11×10^4	11.12×10^4
0.96	10.832	−2.663	22×10^4	22.65×10^4
0.94	11.297	−2.384	35×10^4	34.75×10^4
0.92	11.631	−2.184	47×10^4	47.32×10^4
0.90	11.898	−2.024	60×10^4	60.53×10^4
0.80	12.792	−1.451	1.2×10^6	1.27×10^6
0.70	13.489	−0.907	1.8×10^6	1.80×10^6
0.60	14.144	−0.393	2.5×10^6	2.50×10^6
0.50	14.788	+0.113	3.4×10^6	3.43×10^6
0.40	15.432	+0.616	4.7×10^6	4.74×10^6
0.30	16.072	+1.109	6.6×10^6	6.65×10^6
0.20	16.667	+1.561	9.1×10^6	9.35×10^6
0.10	17.135	+1.932	12×10^6	12.65×10^6
0.00	17.351	+2.128	12×10^6	14.62×10^6

Table VI inspires the following comments:

(1) Near the surface ($r'=1$), Equation (16) is indeterminate, since $P=0$ and $\varrho=0$. But we know by direct observation that T at the surface of the Sun is of the order of 0.5×10^4 degrees. Thus we observe a very *abrupt increase in the temperature near the surface*, as soon as we go from $r'=1.00$ to $r'=0.98$. At a distance from the surface equivalent to $\Delta r'=0.02$, the temperature already reaches 111 200 K.

(2) As soon as we leave the immediate neighbourhood of the surface, the variation of T becomes, on the contrary, very slow. One can even consider, in the very first approximation, that the temperature in the interior of the Sun is constant and of the order of 10^6 degrees. If we push the approximation a little further, we find that the variation of T is *exponential*, but with a very small 'rate of variation': a graphical representation of $\log_e T$ shows that T varies as $\exp(-3.3r')$ over all of the vast range from $r'=0.8$ to $r'=0.1$.

(3) The approximate temperature T_{app} calculated with $X=X_0=0.744$ is correct down to $r'=0.3$. Afterwards, when we come even closer to the center, it still gives a completely reasonable order of magnitude.

(4) Finally, if we compute $P_{rad}=\frac{1}{3}aT^4$ using the exact values of T, we obtain the following values for P_{rad} and for the ratio P_{rad}/P_{gas}. We shall confine Table VII to a few typical values.

TABLE VII

The radiation pressure P_{rad} and the ratio P_{rad}/P_{gas}

r'	T_{exact}	P_{rad}	P_{rad}/P_{gas}
0.98	1.1×10^5	4.4×10^5	0.4×10^{-4}
0.90	6.1×10^5	3.4×10^8	4.3×10^{-4}
0.50	3.4×10^6	3.5×10^{11}	5.7×10^{-4}
0.30	6.7×10^6	4.9×10^{12}	4.1×10^{-4}
0.00	14.6×10^6	11.5×10^{13}	5.2×10^{-4}

5. Summary. The Empirical Representation of the Functions $g(r')$, $\varrho(r')$, $P(r')$, and $T(r')$. The Polytropic Index n

Let us consider the gravitational field strength in the interior of the Sun, given by GM_r/r^2. It can be expressed in terms of the value g_s of the field strength at the surface and of the variables M' and r' by the formula

$$g = g_s M'/r'^2,$$

with

$$g_s = G(M_\odot/R_\odot^2) = 2.74 \times 10^4 \text{ c.g.s.} \tag{17}$$

Knowing M' as a function of r' (Table II), we can calculate the napierian logarithm $\log_e g$ of the gravitational field strength; and knowing ϱ, P, and T as functions of r' (Tables II, III, IV, and VI), we can likewise calculate $\log_e \varrho$, $\log_e P$, and $\log_e T$.

Thus we obtain Table VIII, whose contents are shown graphically in Figure 12;

TABLE VIII

$\log_e g$, $\log_e \varrho$, $\log_e P$ and $\log_e T$ as functions of r'

r'	$\log_e g$	$\log_e \varrho$	$\log_e P$	$\log_e T$
0.90	10.43	−4.66	27.40	13.31
0.80	10.66	−3.34	29.45	14.05
0.70	10.92	−2.09	31.06	14.40
0.60	11.21	−0.90	32.57	14.72
0.50	11.52	+0.26	34.05	15.05
0.40	11.85	+1.42	35.53	15.37
0.30	12.16	+2.55	37.01	15.71
0.20	12.36	+3.59	38.38	16.05
0.10	12.16	+4.45	39.46	16.35
0.00	−∞	+4.90	39.95	16.50

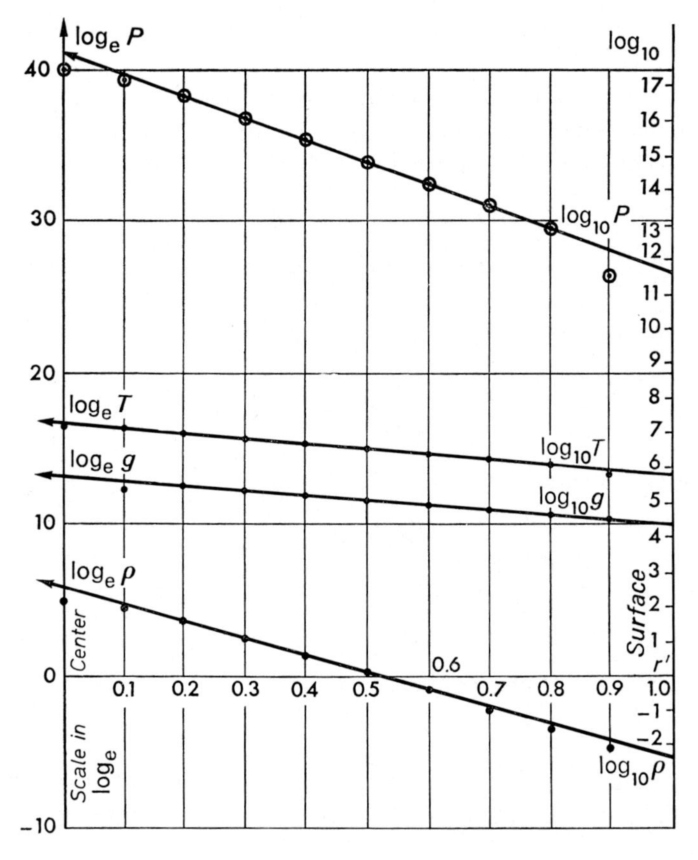

Fig. 12.

in this figure, the ordinate scale on the left corresponds to napierian logarithms, while the ordinate scale on the right corresponds to logarithms to the base 10.

As we see in Figure 12, the logarithms of the functions $\varrho(r')$, $g(r')$, $P(r')$, and $T(r')$ *vary linearly* with r' everywhere except in the immediate neighbourhood of the center and of the surface. The variations of g and T are the slowest; the lines representing them have slopes equal respectively to (-3.10) and (-3.27). The variation of $\log_e \varrho$ is more rapid, with a slope of (-11.4). It is $\log_e P$ that varies most quickly, with a slope of (-14.7). Measuring the 'intercepts', we obtain the c.g.s. formulae

$$\begin{aligned} g &= 4.36 \times 10^5\, \mathrm{e}^{-3.10r'} & \varrho &= 331\, \mathrm{e}^{-11.4r'} \\ P &= 891 \times 10^{15}\, \mathrm{e}^{-14.7r'} & T &= 20 \times 10^6\, \mathrm{e}^{-3.27r'}. \end{aligned} \tag{18}$$

We must take care not to confuse the coefficients of the exponentials with the exact values for $r'=0$, for these empirical representations are not valid for r' less than 0.15, or greater than 0.85.

The fact that it is possible to represent the functions $g(r')$, $\varrho(r')$, $P(r')$, and $T(r')$ by exponentials of the form (18) means that the *logarithmic derivatives* of these functions with respect to r' are *constants*, which in the first approximation are equal to (-3.10), (-11.4), (-14.7), and (-3.27), respectively.

Of course, these constants are not completely independent, and the relations among them can be explained by the *general* relations connecting the functions g, ϱ, P, and T. Two of these relations are especially obvious. Indeed, according to (16) we should have

$$\mathrm{d}T/T = \mathrm{d}P/P - \mathrm{d}\varrho/\varrho\,. \tag{19}$$

Now, according to (18), the terms in this equation are proportional to (-3.27), (-14.7), and (-11.4), respectively. Moreover, if we denote by α the absolute value $(+14.7)$ of the coefficient of r' in the exponential representation of $P(r')$, we have

$$-\frac{\mathrm{d}P}{\mathrm{d}r} = -\frac{1}{R_\odot}\frac{\mathrm{d}P}{\mathrm{d}r'} = +\frac{1}{R_\odot}\alpha P\,, \tag{20}$$

whence, according to (2),

$$P = \frac{R_\odot}{\alpha}\left(-\frac{\mathrm{d}P}{\mathrm{d}r}\right) = \frac{R_\odot}{\alpha}\left(\frac{GM_r}{r^2}\varrho\right) = \text{const}\, g\varrho\,,$$

or

$$\mathrm{d}P/P = \mathrm{d}g/g + \mathrm{d}\varrho/\varrho\,, \tag{21}$$

which corresponds to the approximate formula

$$14.7 \approx 3.1 + 11.4\,.$$

The approximately constant ratio between $(\mathrm{d}\varrho/\varrho)$ and $(\mathrm{d}T/T)$ is called the *polytropic index* and is generally denoted by n, whence the definition

$$n = \frac{(\mathrm{d}\varrho/\varrho)}{(\mathrm{d}T/T)}. \tag{22}$$

In our model of the Sun, n is equal (far from the center and from the surface) to (11.4/3.27) – that is, to 3.5. Using (19), we can also define the polytropic index n by

$$n + 1 = \frac{(\mathrm{d}P/P)}{(\mathrm{d}T/T)}. \tag{23}$$

Models in which n is assumed to be constant *throughout the star* have played a very important historical role in the development of the theory of the internal structure of stars (cf. [6], for example).

6. The (Superficial) 'Convective Zone' of the Sun

Starting from the photosphere and going downwards into the star, we rapidly encounter a region where the temperature is relatively very high. It is so high that hydrogen (principal component of the 'stellar mixture') is already *partially ionized.*

Let us consider a volume element $(\mathrm{d}V)$ formed of pure H, partially ionized. Let us assume that it is raised towards the surface (where T is lower than in the region from which it came). In becoming colder, the mixture of ions and free electrons (analogous to a 'liquid-vapour' mixture) will tend to 'condense' – that is, to *recombine* into *neutral* atoms. But just as the condensation of a vapour releases heat, the recombination of ions is accompanied by a *release of energy* (the *energy* of the photons emitted during the recombination, or the energy transferred to the free electrons during 'collisions of the second kind'). This release of energy has the effect of heating the element $(\mathrm{d}V)$.

Thus the element $(\mathrm{d}V)$ in cooled *less* than is required for thermal equilibrium with its surroundings. But since it remains hotter and the outside pressure remains unchanged, the element $(\mathrm{d}V)$ takes on a density ϱ' different from ϱ, the density of the elements which normally occupy the arrival 'level', and which surround $(\mathrm{d}V)$ upon its arrival at this level. Indeed, $\varrho' = (P/T')(\mu m_{\mathrm{H}}/k)$, while $\varrho = (P/T)(\mu m_{\mathrm{H}}/k)$. Since the temperature T' of $(\mathrm{d}V)$ at its arrival is greater than the temperature of the surrounding elements, while P remains unchanged (and μ, in the first approximation, also remains the same), ϱ' *will be smaller than* ϱ. Consequently $(\mathrm{d}V)$ will continue to rise instead of returning to its starting point.

But 'higher up', there will be *fewer ions* in the element $(\mathrm{d}V)$ (which is *cooled* less than is required for equilibrium with its surroundings, but which is *cooled* all the same), and *recombination will be less important.* Fairly rapidly, $(\mathrm{d}V)$ will reach a level at which the physical process of 'heating by recombination' will be stopped: $(\mathrm{d}V)$ will not go all the way to the surface of the Sun.

Now, the reverse process will take place for the elements $(\mathrm{d}V')$ which start from a relatively high level (the one at which $(\mathrm{d}V)$ stopped in its ascent) and begin to descend under the effect of some perturbation. Because of the descent, there will be more ionizations. The ionizations will 'consume energy' and the descent will continue; for in becoming colder than the surrounding elements, $(\mathrm{d}V')$ will have a higher density than these 'normal' elements, and the 'buoyant force' will not be sufficient to stop the fall of (dV'). But just like the rise of (dV), the descent of (dV') will not be unlimited. It will be stopped as soon as all the hydrogen in (dV') is *completely* ionized.

The rising elements (dV) and the descending elements (dV') produce a *circulation* of the matter in a certain *convective 'layer'* or *'zone'*. In this 'convective zone', the temperature of the moving elements 'contradicts', so to speak, the ambient temperature (the cold layers are heated and the hot layers are cooled), tending to equalize the temperature in the different parts of the 'convective zone'. In other words, the temperature gradient in the 'convective zone' should be *smaller* than it would have been without convection. In the Sun, the convective region corresponds to

$$r' > 0.86 \quad (\Delta r \approx 100\,000 \text{ km}).$$

The condition for the existence of a convective zone can be written ([1], p. 46) (the condition for *instability*):

$$\left(\frac{d \log T}{d \log P}\right)_{\text{conv}} \leqslant \frac{d \log T}{d \log P}. \tag{24}$$

It can be shown (see Schwarzschild's book, pp. 47–52) that by a fortunate coincidence (which is not completely accidental, since it is the result of a number of simplifying hypotheses) the relation between the total pressure P and the volume V of an element of the convective zone is of the same form as the relation describing the adiabatic expansion of a perfect gas:

$$PV^{\gamma} = \text{const}.$$

For a monatomic gas, like hydrogen at a high enough temperature (to prevent the formation of H_2 molecules), $\gamma = \frac{5}{3}$.

This entails

$$dP/P + \gamma (dV/V) = 0.$$

But since a perfect gas also obeys the relation $PV = (k/m_H)\,T$, we have

$$dP/P + dV/V = dT/T.$$

Whence, eliminating dV/V,

$$\frac{dT}{T} = \frac{\gamma - 1}{\gamma} \frac{dP}{P},$$

or, dividing by dr' (and going to logarithms to the base 10),

$$\frac{d \log_{10} T}{dr'} = \frac{\gamma - 1}{\gamma} \frac{d \log_{10} P}{dr'}. \tag{25}$$

To 'test' our $\varrho(r')$ in the convective region $(r' > 0.86)$, we can use the general relations

$$dP/dr' = -\,1.91 \times 10^{15} \frac{M'\varrho}{r'^2} \quad \text{and} \quad T/P = 7.2 \times 10^{-9} \frac{1}{\varrho},$$

together with the 'convective' law (for $\gamma = \frac{5}{3}$) above; the latter can also be written

$$\frac{dT}{dr'} = \frac{2}{5}\frac{T}{P}\frac{dP}{dr'} \quad \text{or} \quad \left(\frac{d\log T}{d\log P}\right)_{\text{conv}} = 0.4. \tag{26}$$

Replacing dP/dr' and T/P by their numerical values, we have

$$(dT/dr')_{\text{conv}} = -\,5.50 \times 10^6 \frac{1}{r'^2} \tag{27}$$

(we have replaced M' by 1, for in this region M' is equal to 1 to within 1%).

The 'test' can also be put in the following, especially convenient form:*

$$r'^2 (\Delta T)_{\Delta r' = -0.02} = +\,11.0 \times 10^4 = \text{const}. \tag{28}$$

We should not be surprised that $\varrho(r')$ is tested by means of a relation in which ϱ does not appear *explicitly*, for the values of ΔT in the last equation are deduced from $T(r')$, which was derived from $P(r')$, which in turn was calculated from a certain density distribution. Table IX gives the numerical values, which confirm the validity of the test (the 'discrepancies' are due to having taken too large a value of $\Delta r'$).

We have just shown how a certain choice of the density distribution $\varrho(r')$ determines the functions $g(r')$, $P(r')$, and $T(r')$.

TABLE IX

r'	0.98	0.97	0.96	0.95	0.94	0.93	0.92	0.91	0.90
$T_{ex} \times 10^{-4}$	11.12		22.65		34.75		47.32		60.53
$\Delta T_{ex} \times 10^{-4}$		11.03		12.10		12.57		13.21	
$r'^2 \times \Delta T_{ex} \times 10^{-4}$		10.9		10.9		10.9		10.9	

We shall now turn to the properties which make it possible to choose the 'right' distribution $\varrho(r')$, taking into account not only the total mass and the radius of the Sun, but also its total luminosity $L_\odot = 3.90 \times 10^{33}$ ergs sec^{-1} and the influence of its chemical composition on the generation of energy by thermonuclear reactions.

Exercise 3

(1) Review the definition of the 'gravitational potential' $\varphi(r)$. Show that if we take $\varphi(\infty) = 0$, we have

$$\varphi(R) = -\,\varphi_0 = -\,GM/R,$$

provided that we are dealing with a star that satisfies a simple condition (of structure). Recall this condition. (Take care not to confuse $\varphi_0 = GM/R$ with $\varphi(0)$.)

(2) Show that the 'escape velocity' at the surface of the star, $v_e(R)$, is given by $v_e(R) = (2\varphi_0)^{1/2}$. ('Escape' = ability to recede to infinity.)

* Of course, we can also integrate (27) analytically and find $T = 5.50 \times 10^6 (1/r' - 1)$, which is also easy to test.

(3) Let $\bar{\varrho}$ be the mean density of a star. We set

$$\varrho' = \varrho/\bar{\varrho}; \quad P' = P/P_0; \quad T' = T/T_0;$$
$$\beta = P_{\text{gas}}/P; \quad \eta = \beta\mu m_{\text{H}}/k.$$

Show that by an appropriate choice of the functions $P_0 = f(\bar{\varrho}, \varphi_0)$ and $T_0 = \Theta(\eta, \varphi_0)$, where f and Θ do not depend explicitly on r (but may depend on r through η), the equations of mechanical equilibrium of the star take on the simple form:

$$\mathrm{d}M'/\mathrm{d}r' = (P'/T')\, r'^2 \tag{1}$$

and

$$\mathrm{d}P'/\mathrm{d}r' = -\, 3M'\varrho'/r'^2. \tag{2}$$

Find the functions f and Θ, and show that we always have $P'/T' = 3\varrho'$.

(4) To the real model of question (3) we fit a 'theoretical' model determined by the condition $\varrho' = 1$ (for all values of r'). Integrate the system of Equations (1) and (2) under this hypothesis, with the usual boundary conditions.

(5) Let $P_c{}^h$ and $T_c{}^h$ be the central pressure and the central temperature of the 'homogeneous' model (h) of question (4). Does the 'homogeneous' model (h) give a relatively simple physical interpretation of the quantities P_0 and T_0 introduced in question (3)?

Exercise 4

We recall that according to one of Schwarzschild's models of the Sun, the distributions of the pressure P, the density ϱ, the temperature T, and the ratio M'/r'^2 (where $M' = M_r/M_\odot$) are given as functions of the distance $r' = r/R$ from the center by the following table:

r'	0.0	0.1	0.2	0.3	0.4	0.5	0.6	0.7	0.8	0.9
$\log P$	39.95	39.46	38.38	37.01	35.53	34.05	32.57	31.06	29.45	27.40
$\log \varrho$	+4.90	+4.45	+3.59	+2.55	+1.42	+0.26	−0.90	−2.09	−3.34	−4.66
$\log T$	16.50	16.35	16.05	15.71	15.37	15.05	14.72	14.40	14.05	13.31
$\log(M'/r'^2)$	$-\infty$	1.94	2.14	1.94	1.63	1.30	0.99	0.70	0.44	0.209
M'	0.00	0.07	0.34	0.63	0.82	0.92	0.967	0.988	0.996	0.999

(1a) Show by means of a graph that for $0.1 < r' < 0.9$ the distribution of P, ϱ, T, and M'/r'^2 can be represented in the following form, where α, β, τ, and τ' are positive constants:

$$P = P'_c \exp(-\,\alpha r'), \tag{1}$$
$$\varrho = \varrho'_c \exp(-\,\beta r'), \tag{2}$$
$$T = T'_c \exp(-\tau r'), \tag{3}$$
$$M'/r'^2 = A_0 \exp(-\,\tau' r'). \tag{4}$$

Use a scale such that 2 cm in the abscissa represents $\Delta r' = 0.1$, and 1 cm in the ordinate represents $\Delta \log_e F = 2$ (where F is P, ϱ, T, or M'/r'^2).

(1b) Compute the value of α, β, τ, τ', P'_c, ϱ'_c, T'_c and A_0, using the method of least squares. Graph the solutions obtained in this way.

(1c) What simple relation exists between τ and τ' in the first approximation? What is the approximate expression for τ in terms of α and β?

(2) Using the general formulae of internal structure, justify the relation between τ, α, and β found in question (1), and establish a relation for μ in terms of T'_c, ϱ'_c, and P'_c.

According to this relation, find the value of μ corresponding to the values of T'_c, P'_c, and ϱ'_c obtained in question (1). Is this solution reasonable?

(3) Give a theoretical justification for the relation between τ and τ' found in question (1). Find a theoretical expression for A_0 in terms of α, P'_c, and ϱ'_c.

(4) Calculate the numerical value of A_0 from the relation found in question (3). Account for the disagreement between this value and the value found in question (1).

(5) Using the value of A_0 found in question (4) and the value of τ' derived from α and β, show that the formula $M' = A_0 r'^2 \exp(-\tau' r')$ closely represents the value of M' for $r' = 0.7$ and $r' = 0.3$, but that it is less satisfactory for $r' = 1.0$, although it gives a reasonable order of magnitude.

(6) Prove by a simple theoretical argument that the formula $M'/r'^2 = A_0 \exp(-\tau' r')$ is certainly unacceptable in the vicinity of $r' = 0$.

(7) Show that according to the general equations of internal structure the single hypothesis $P = P'_c \exp(-\alpha r')$, where P'_c and α are constants, implies the possibility of finding $(M')^2$ as a function of r'. Calculate $(M')^2$ explicitly in terms of r' and α by performing the necessary integrations. (Let C be the quantity $48 \times 2.13\ P'_c/1.91 \times 10^{15}$.)

(8) Let $\varphi(x)$ be the expansion of e^x up to and including the term in x^4. Calculate $\varphi(\alpha)$.

(9) Apply the expression for $(M')^2$ found in question (7) to $r' = 1$, and deduce an approximate value of P'_c assuming $\alpha = 14.87$. Compare the results with those obtained in question (1).

(10) Assuming $C/\alpha^4 = 1$ and $\alpha = 14.87$, calculate $\varphi(\alpha r')$ for $r' = 0.3$, 0.4, 0.5, 0.6, 0.7, 0.8 and 0.9. Show that the formula for M' found in question (7) gives an excellent approximation to M' for all these values of r'.

(11) Does the formula found in question (7) and checked in question (10) seem to you to be compatible with the empirical expression for M'/r'^2 obtained in question (1)?

Exercise 5

(1) We assume that the Sun is composed of pure hydrogen and that the temperature in the interior is independent of $r' = r/R$ (where r is the distance from the center and R is the solar radius). In addition, we assume that the mean mass per particle μ is independent of r'. Finally, we assume that the total pressure varies as a function of r' according to a law of the form

$$P = P'_c e^{-\alpha r'} \quad \text{with} \quad \alpha = 14.7 \quad (R = 6.96 \times 10^{10} \text{ c.g.s.}). \tag{1}$$

N.B. – Avoid using the equation in dP/dr throughout this problem!

(1.1) Discuss these hypotheses in the light of your knowledge of the 'exact' solutions of the problem of the internal structure of the Sun. (Use the concepts of a 'rough' approximation and a 'reasonable' hypothesis.)

(1.2) Using the above hypotheses, calculate expression (2) for $M' = M_r/M$ in terms of α, r', and the central density ϱ'_c associated with P'_c ($M = 1.99 \times 10^{33}$ c.g.s.).

(1.3) Show that to a very good approximation, expression (2) can be replaced at $r' = 1$ by a much simpler expression (3), which has only one term in α (to the third power).

(1.4) Write (3) and derive from it a relation between α^3 and ϱ'_c.

(1.5) Compute the corresponding numerical value of ϱ'_c.

(2) We retain all the preceding hypotheses except $T = \text{const}$. This latter assumption is replaced by a more correct, but still relatively simple, hypothesis concerning the variation of T.

(2.1) Explain why the value ϱ'_c found above is markedly higher than the exact central density ϱ_c.

(2.2) Using this more correct function $T(r')$, find a more accurate value for ϱ'_c. Let ϱ''_c be this value.

(2.3) Considering the expression for ϱ''_c in terms of μ, P'_c, and T'_c (the central temperature of the function $T(r')$), explain why ϱ''_c is still too high in comparison with ϱ_c.

Exercise 6

We propose to study a model of a *very massive*, spherically symmetric star, in which the ratio P_g/P of the gas pressure P_g to the total pressure P has the same constant value β at all points. We denote the radiation pressure by P_r, and the mean mass of a particle of the mixture in units of H (where $H = m_H$ is the mass in grams of a hydrogen atom) by μ. We assume that μ has the same value throughout the star. We recall that

$$c = 3 \times 10^{10} \text{ c.g.s.} \qquad a = 7.6 \times 10^{-15} \text{ c.g.s.}$$
$$G = 6.67 \times 10^{-8} \text{ c.g.s.} \qquad k/H = 8.32 \times 10^{7} \text{ c.g.s.}$$

(I.1) Let ϱ_g be the gas density at a point in the star. Recall the expressions for P_g and P_r in terms of the temperature T. Put P in the form $K(\beta)\varrho_g{}^n$, where $K(\beta)$ depends only on β, μ, and the constants a, k, and H, while n is a numerical exponent to be determined.

(I.2) Let E be the *density* of the total energy at a point in the star (the same point as in the preceding question), including the density of the 'rest' energy E_g (which corresponds to the mass, according to Einstein's mass-energy equation), the density of the thermal energy of the gas E_k, and the density of the radiation energy E_r. Give expressions for E_g in terms of ϱ_g, E_r in terms of P_r, and E_k in terms of P_g. (Treat the mixture as a perfect monatomic gas.) Give E in terms of $K(\beta)$, ϱ_g, and β.

(II) We assume that the equations of mechanical equilibrium for each point of a *very massive* star have the following form, which is different from the usual form:

$$\mathrm{d}M(r)/\mathrm{d}r = 4\pi r^2 E/c^2, \tag{I}$$

$$[(1-2GM(r)\,c^{-2}r^{-1})/(P+E)]\,r^2\,\mathrm{d}P/\mathrm{d}r + GM(r)/c^2 + (4\pi G/c^4)\,r^3 P = 0, \tag{II}$$

where r is analogous to the real distance from the center (which in this case we shall call $\bar{r}$). The variables r and $\bar{r}$ are related by the formula

$$\bar{r} = \int_0^r [1-2GM(r)/c^2 r]^{-1/2}\,\mathrm{d}r. \tag{III}$$

We assume that r reduces to $\bar{r}$ in the 'classical' case.

(II.1) What are the equations corresponding to (I) and (II) in the 'classical' case of small mass? What is the *name* of the theory which enables us to establish Equations (I), (II), and (III)?

(II.2) We replace the variable ϱ_g by a new variable θ defined by $\varrho_g = \varrho_{gc}\theta^3$, where ϱ_{gc} denotes the value of ϱ_g at the center of the star, and we put

$$\alpha = K(\beta)\varrho_{gc}{}^{1/3}/c^2 \quad \text{and} \quad f = 4 - \tfrac{3}{2}\beta.$$

Find expressions (1′) for P and (2′) for E in terms of c, α, θ, ϱ_{gc} and f.

(II.3) We introduce the dimensionless variables ξ and $v(\xi)$ defined by the relations $r = \lambda\xi$ and $M(r) = 4\pi\lambda^3\varrho_{gc}v(\xi)$, where λ is a certain *length* which depends (Equation (IV)) on ϱ_{gc}, α, and some universal constants. What are the values of θ and $v(\xi)$ for $\xi = 0$? Find an expression (I′) for $\mathrm{d}v/\mathrm{d}\xi$ in terms of ξ and θ (where α and f enter as parameters). Show that for a certain value of λ given by (IV), the differential Equation (II′) satisfied by the function $\theta(\xi)$ can be written in the form

$$[(1-8\alpha v\xi^{-1})/(1+f\alpha\theta)]\,\xi^2\,\mathrm{d}\theta/\mathrm{d}\xi + v + \alpha\xi^3\theta^4 = 0. \tag{II′}$$

Find Equation (IV) for λ. Specify the mathematical nature of the differential system formed by (I′) in conjunction with (II′). Let ξ_1 be the smallest value of ξ such that $\theta(\xi_1) = 0$. What is the physical meaning of ξ_1? Do the values of ξ_1 and $v(\xi_1)$ necessarily depend on α and β?

(III) (*N.B.* – The 'numerical coefficients' appearing in the equations required in questions (3), (4), and (5) below are to be calculated to two significant figures, using a slide rule.)

(III.1) Let T_c be the central temperature and P_c the total central pressure. Express T/T_c and P/P_c in terms of θ.

(III.2) Let ϱ be the *total* density (corresponding to E) and let ϱ_c be the 'central' value of ϱ. Express ϱ/ϱ_c and ϱ_g/ϱ_c in terms of θ (with α and f as parameters).

(III.3) Find an expression for the total mass M of the star in units of the solar mass $M_\odot$, as a function of β, μ, and $v(\xi_1)$ alone. (All the universal constants should be combined with the numerical factors into a single 'numerical coefficient', to be calculated.)

(III.4) Similarly, find expressions for T_c, ϱ_{gc}, and P_c in terms of α, β, and μ alone. This time, use c.g.s. units.

(III.5) We associate with ξ the variable $\bar{\xi}$, defined by $\bar{r} = \lambda\bar{\xi}$. The real radius of the star is then $\bar{R} = \lambda\bar{\xi}_1$, where $\bar{\xi}_1 = \bar{\xi}(\xi_1)$. Find an expression for $\bar{R}$ in units of the solar radius $R_\odot$, as a function of α, β, μ, and $\bar{\xi}_1$. Calculate the corresponding 'numerical coefficient'.

(IV) Throughout this section we consider the case in which β is much less than 1 in a very massive star.

(IV.1) In this case, what is the predominant term in the total pressure?

(IV.2) Write Equations (I″) and (II″) which now replace (I′) and (II′), taking $f = 4$. Do the quantities $\bar{\xi}_1$ and $v(\xi_1)$ still depend on β?

(IV.3) Show that, when the results obtained in Section III, question (3), are taken into account, the masses of the corresponding stars are all much greater than the mass of the Sun.

(IV.4) Give the expression for M in terms of β, μ, and $v(\xi_1)$, and the expression for $\bar{R}$ in terms of α, β, μ, and $\bar{\xi}_1$, obtained by replacing $(1-\beta)$ by 1. If we assumed $\bar{\xi}_1$ and $v(\xi_1)$ to be independent of α, what would be the form of the curves $M=$ const. and $\bar{R}=$ const. on a graph with $\log\alpha$ as abscissa and $\log\beta$ as ordinate?

(IV.5) We give in the following table part of the results from the numerical solution of Equations (I″) and (II″):

$\log\alpha$	$\bar{\xi}_1$	$v(\xi_1)$
-4.0	6.8975	2.0158
-3.8	6.8980	2.0144
-3.6	6.8986	2.0122
-3.4	6.8998	2.0087
-3.2	6.9013	2.0032
-3.0	6.9040	1.9945
-2.8	6.9083	1.9809
-2.6	6.9153	1.9596
-2.4	6.9267	1.9266
-2.2	6.9456	1.8763
-2.0	6.9780	1.8008

On the other hand, the results of the numerical solution of the more general Equations (I′) and (II′) for the masses and radii of pure-hydrogen models can be represented in the $(\log\alpha, \log\beta)$ plane by the graph below (Figure A). Explain the form of the curves $M=$ const. and $\bar{R}=$ const. in the region defined by $\log\alpha \leqslant -2.0$ and $\log\beta \leqslant -1.0$.

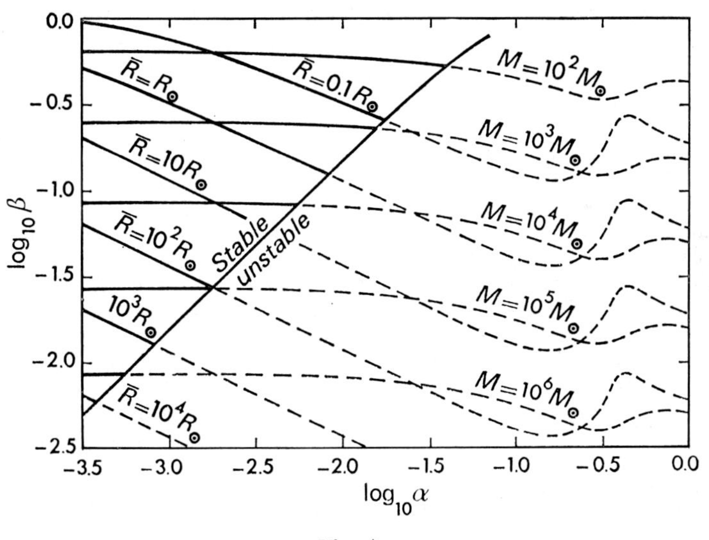

Fig. A.

(IV.6) A study of the *dynamical* stability of the model leads to a separation of the graph into two regions. For $\beta \ll 1$, the boundary between the regions of stability and instability can be represented by the straight line whose equation is $\beta = 15.795\alpha$. Find approximate expressions for $\bar{R}$ and ϱ_{gc} in terms of M and μ for a high-mass model in the neighbourhood of the stability limit, taking $\bar{\xi}_1 = 6.90$ and $v(\xi_1) = 2.02$. Compute $\bar{R}$ and ϱ_{gc} for a pure-hydrogen model with the following values of M: 10^6, 10^7, 10^8, and 10^9 $M_\odot$.

(IV.7) To what kind of astronomical objects might we be inclined to apply these models?

Exercise 7

We consider a model of a star consisting of a gaseous sphere which is chemically homogeneous and in equilibrium, and in which the total pressure P at each point is related to the corresponding density by the equation

$$P = k\varrho^{\gamma}, \tag{1}$$

where k is a constant characteristic of the star and γ is a constant which is the same for all stars and whose numerical value will be given below. We assume that for a given star, the physical state at any point depends only on the distance r from the center.

(1) Let $M_r(r)$ be the mass contained in a sphere of radius r. Write the equation of mechanical equilibrium, which we shall call (2).

(2) Write an expression (3) for the mass $\mathrm{d}M_r$ contained in a spherical layer of thickness $\mathrm{d}r$.

(3) Write the second-order differential equation (4) obtained by eliminating M_r between (2) and (3). Explain why (1) and (4) make it possible to find the internal structure of the star.

(4) Let ϱ_0 be the density at the center of the star. We put

$$\varrho(r) = \varrho_0\theta^n, \quad \text{with} \quad n = 1/(\gamma - 1) \quad \text{and} \quad r = \alpha\xi,$$

where α represents a characteristic length defined by the equation

$$\alpha^2 = k\gamma n(4\pi G)^{-1}\, \varrho_0^{\gamma-2},$$

in which G is the constant of gravitation.

(4a) What are the values of ξ, θ, and $\mathrm{d}\theta/\mathrm{d}\xi$ at the center of the star?

(4b) Find the second-order differential Equation (4′) satisfied by the function $\theta(\xi)$.

(5) *In all the rest of the problem* we adopt the value $\gamma = 1.2 = 6/5$.

(5a) What is the value of the polytropic index of the corresponding model? What is the index of Eddington's standard model?

(5b) Show that the function

$$\theta(\xi) = (1 + \tfrac{1}{3}\xi^2)^{-1/2}$$

correctly describes the equilibrium structure of the model, when we take suitable boundary conditions at the center of the star. Indicate the general appearance of the graphical representation of this function $\theta(\xi)$.

(6a) Explain why in the present case it is not possible to define the 'surface' of the star by setting the surface density at this point equal to zero.

(6b) Let ξ_s be the value of ξ at the surface of the star. We shall define ξ_s by expressing the fact that the total optical depth for $\xi_s < \xi < \infty$ is equal to 1. Calculate ξ_s in terms of $\varrho_0\alpha$ and the absorption coefficient per *gram* k_0, making the simplifying hypotheses $\xi_s \gg 1$ and $k_0 = \text{const.}$, and taking care not to confuse k_0 with the absorption coefficient in cm^2 per cm^3 (cf. [9], Chapter VII).

We shall use these hypotheses throughout the rest of the problem.

(7) Write the perfect gas law in terms of the density ϱ, the temperature T and mean molecular mass μ. We recall that μ stands for the mean mass of a particle in terms of unit atomic weight ($1.659\,75 \times 10^{-24}$ g).

We assume that the perfect gas law can be applied to all parts of the star and that radiation pressure is negligible. Express the temperature T_s at the surface of the star in terms of the central temperature T_0 and of $\theta_s = \theta(\xi_s)$.

(8) Express the radius R of the star in terms of ϱ_0, k, and the constants k_0 and G. (It is not necessary to compute the numerical value of the product $2^{-9/8}3^{5/4}$, which can be denoted by λ_0.)

(9) Express the total mass M of the star in terms of ϱ_0, k, and G. (It is not necessary to compute a numerical value for $9\sqrt{2}/\sqrt{\pi}$.)

(10) Express the luminosity L of the star in terms of ϱ_0, k, μ, k_0, G, R_g, and σ (where R_g is the gas constant and σ is the constant in Stefan's law). We shall assume that the star radiates as a black body at the temperature T_s.

(11a) Show that all the stars described by the model under consideration satisfy a mass-luminosity-radius relation of the form $L = AR^pM^q$ and calculate the coefficients A, p, and q.

(11b) Can it be said that the mass-luminosity-radius relation obtained is the same for all the stars under consideration?

(12a) Compute A numerically, taking $\mu = 0.6$, $G = 6.7 \times 10^{-8}$, $\sigma = 5.7 \times 10^{-5}$, and $R_g = 8.3 \times 10^7$.

(12b) Compute the value of A from the data for the Sun: $L_\odot = 3.6 \times 10^{33}$, $R_\odot = 7 \times 10^{10}$, $M_\odot = 2 \times 10^{33}$.

Compare the result with that obtained in question (12a). What should we conclude from this comparison?

(13) The observations show that for main-sequence stars there exists a relation of the form $L \propto M^4$ and a relation of the form $R \propto M^{3/4}$ (where $\propto$ means 'proportional to').

(13a) Is the theoretical relation obtained in question (11) in agreement with the observational results?

(13b) Discuss the reasons which might explain the result of question (13a).

CHAPTER IV

ENERGY EQUILIBRIUM AND NUCLEAR REACTIONS

1. The Equation of Energy Equilibrium

Let $L_r = L_r(r)$ be the function representing the *net* flow of integrated radiation (cf. [9], Chapter II, Section 2 and Chapter VII, Section 5) across a sphere (Σ_r) of radius r.

L_r obviously remains *constant* (and equal to its value L at the surface) as long as we remain outside the central region where nuclear reactions take place. In the present state of evolution of the Sun, this region corresponds to about $r' \leqslant 0.3$.

Let us now consider a layer $(r, \mathrm{d}r)$ located at some distance r from the center. The energy equilibrium *in* this layer $(r, \mathrm{d}r)$ requires that the *net outflow of energy per second* crossing the boundaries of $(r, \mathrm{d}r)$ be equal to any '*energy generation*' (by nuclear reactions) in $(r, \mathrm{d}r)$. More explicitly, let ε_r be the average energy produced per second by nuclear reactions in each *gram* of the 'stellar mixture' making up $(r, \mathrm{d}r)$. The energy produced per second in each cm^3 of $(r, \mathrm{d}r)$ will be given by $\varepsilon_r \varrho$, where ϱ is the density at a distance r from the center.

The energy produced per second throughout $(r, \mathrm{d}r)$ will be

$$(4\pi r^2 \, \mathrm{d}r) \, \varepsilon_r \varrho = \varepsilon_r \, \mathrm{d}M_r .$$

The balance between 'gains' and 'losses' per second then requires that we have (using obvious notation):

$$\varepsilon_r \, \mathrm{d}M_r + L_r^+ + L_{(r+\mathrm{d}r)}^- = L_r^- + L_{(r+\mathrm{d}r)}^+ .$$

Or, by analogy with Equation (12) of Section II.4 of [9]:

$$\varepsilon_r \, \mathrm{d}M_r = [L_{(r+\mathrm{d}r)}^+ - L_{(r+\mathrm{d}r)}^-] - [L_r^+ - L_r^-] = L_{(r+\mathrm{d}r)} - L_r .$$

Dividing by $\mathrm{d}r$ and letting $\mathrm{d}r$ approach zero, we obtain:

$$\varepsilon_r \, \mathrm{d}M_r/\mathrm{d}r = \mathrm{d}L_r/\mathrm{d}r .$$

Thus we find that

$$\boxed{\mathrm{d}L_r/\mathrm{d}r = 4\pi r^2 \varepsilon_r \varrho .}$$

In the regions (r' greater than about 0.3) where $\varepsilon_r = 0$, energy equilibrium implies that $\mathrm{d}L_r/\mathrm{d}r = 0$, and we recover $L_r = \text{const}$.

Remark. The details of the calculation ε_r will be given in Section 3. For the moment we shall limit ourselves to a few generalities.

2. The p-p Chain and the C-N Cycle

2.1. Introduction

The energy generation per gram per second, ε_r, depends on the composition, the temperature T, and the density ϱ of the stellar mixture at a distance r from the center.

We shall find (see Section 3) that for temperatures of the order of 10^6 degrees (to be exact, between 4×10^6 and 50×10^6 degrees), the only long-term reactions possible under stellar conditions are:

(1) *the p-p chain* (between about 4×10^6 and 25×10^6 degrees);

(2) *the C-N cycle* (between about 12×10^6 and 50×10^6 degrees).

Now the charge Z_H of the proton is 1. That of the carbon nucleus, Z_C, is 6. The electrostatic repulsion at a given distance is weaker for the (p-p) pair than for any other pair of nuclei. It is only 6 times stronger for the (p-C) pair and 7 times stronger for the (p-N) pair. The nuclei with Z less than 6 do not give rise to 'cyclic' reactions and are not abundant enough to react with protons during a long period of time. The nuclei with Z greater than 6 or 7 repel the protons too strongly to get close enough to them, at velocities corresponding to temperatures of the order of 10^6 degrees. They remain too far from the protons for the 'nuclear forces' to come into play.

2.2. An explicit schematic representation of the composition of nuclei

Let us agree to represent the proton p by $\oplus$, the negative electron by β^- or by $\boxed{-}$ the positron β^+ by $\boxed{+}$, the neutron n by $\circ$, the neutrino by $\boxed{\nu}$, and the more complex particles by a rectangle 'filled' with the appropriate components.

We recall that the *name* of a nucleus, in chemistry, depends only on the number of *protons*. The mass of the nucleus, in units of m_H, is given by a superscript. We can then draw up Table X.

Remark. A more logical and clear arrangement of Table X can be seen in Table XI (but it would take up too much space if all the details were to be given).

2.3. The details of the reactions in the p-p chain (Bethe, 1938)

Agreeing to connect by a + sign the reacting nuclei on the one hand, and the 'products' of the reaction on the other hand (to the right of the arrow), we can represent the 'proton-proton' chain proposed by Hans Bethe, in 1938, in the following manner. (We denote by Q_i the energy released in each reaction.)

At the beginning, we have the *fusion* of two pairs of protons, followed by the annihilation of the β^+ and the β^-:

$$\left.\begin{cases} \oplus + \oplus \rightarrow \boxed{\begin{matrix}\circ\\ \oplus\end{matrix}} + \boxed{+} + \boxed{\nu} + \tfrac{1}{2}Q_1 \quad \text{followed by} \quad \boxed{+} + \boxed{-} \rightarrow \tfrac{1}{2}Q_2 \\ \oplus + \oplus \rightarrow \boxed{\begin{matrix}\circ\\ \oplus\end{matrix}} + \boxed{+} + \boxed{\nu} + \tfrac{1}{2}Q_1 \quad \text{followed by} \quad \boxed{+} + \boxed{-} \rightarrow \tfrac{1}{2}Q_2 \end{cases}\right\} . \quad \text{(I)}$$

TABLE X

Z	Normal nucleus	Corresponding isotope	
$Z=1$	p = ⊕ = H^1 Proton	D = [○ ⊕] = H^2 Deuteron	
$Z=2$	α = [○ ○ ⊕ ⊕] = He^4 Alpha particle	[○ ⊕ ⊕] = He^3 Light isotope	
$Z=6$	© = [○ ○ ○ ○ ○ ○ ⊕ ⊕ ⊕ ⊕ ⊕ ⊕] = C^{12}	[© ○] = C^{13} Heavy isotope	
$Z=7$	[© ○ ⊕] = N^{14}	[© ⊕] = N^{13} Light isotope	[© ○ ○ ⊕] = N^{15} Heavy isotope
$Z=8$	[© ○ ○ ⊕ ⊕] = O^{16}	[© ○ ⊕ ⊕] = O^{15} Light isotope	

Remark: ⊕ = ○ + [+]   Proton = neutron + positive electron

TABLE XI

Z \ A	$A=1$	$A=2$	$A=3$	$A=4$	$A=12$	$A=13$	$A=14$	$A=15$	$A=16$
$Z=1$	p = H^1	D = H^2							
$Z=2$	......	......	He^3	He^4					
$Z=6$	......	......	......	......	C^{12}	C^{13}			
$Z=7$	......	......	......	......	....	N^{13}	N^{14}	N^{15}	
$Z=8$	......	......	......	......	....	....	....	O^{15}	O^{16}

Remark. In the presence of a large excess of [−], all of the [+] produced are annihilated. The total energy released is (Q_1+Q_2).

The 'formal' transcription of this reaction is:

$$2(p+p) \rightarrow 2D + 2\beta^+ + 2\nu + Q_1 \quad \text{followed by} \quad 2(\beta^+ + \beta^-) \rightarrow Q_2 . \qquad (1)$$

The chain continues with the pair of reactions:

$$\left\{ \begin{array}{l} [\bigcirc\ \oplus] + \oplus \rightarrow [\bigcirc\ \oplus\ \oplus] + \tfrac{1}{2}Q_3 \\ \ldots\ldots\ldots \text{Ditto} \ldots\ldots\ldots \end{array} \right\} . \qquad \text{(II)}$$

Transcription:

$$2(\mathrm{D}+\mathrm{p}) \rightarrow 2\mathrm{He}^3 + Q_3. \tag{2}$$

And the chain ends with:

$$\boxed{\begin{matrix}\bigcirc & \\ \oplus & \oplus\end{matrix}} + \boxed{\begin{matrix}\bigcirc & \\ \oplus & \oplus\end{matrix}} \rightarrow \boxed{\begin{matrix}\bigcirc & \bigcirc \\ \oplus & \oplus\end{matrix}} + \oplus + \oplus + Q_4. \tag{III}$$

Transcription:

$$\mathrm{He}^3 + \mathrm{He}^3 \rightarrow \mathrm{He}^4 + \mathrm{p} + \mathrm{p} + Q_4. \tag{3}$$

The result of this chain is thus expressed 'physically' by:

$$6\mathrm{p} + 2\beta^- \rightarrow \alpha + 2\mathrm{p} + 2\nu + Q_{\mathrm{pp}}, \tag{4}$$

and 'mathematically' by the equation obtained by considering only the final result ('simplifying' by 2p!):

$$4\mathrm{p} + 2\beta^- \rightarrow \alpha + 2\nu + Q_{\mathrm{pp}}, \tag{5}$$

or

$$4\mathrm{H}^1 + 2\beta^- \rightarrow \mathrm{He}^4 + 2\nu + Q_{\mathrm{pp}}. \tag{5'}$$

Remark. In this last equation, the symbols H^1 and He^4 represent the *nuclei* and not the *atoms* of hydrogen and helium, respectively.

In our schematic notation reaction (5) is written:

$$\oplus + \oplus + \oplus + \oplus + \boxed{-} + \boxed{-} \rightarrow \boxed{\begin{matrix}\bigcirc & \bigcirc \\ \oplus & \oplus\end{matrix}} + \boxed{\nu} + \boxed{\nu} + Q_{\mathrm{pp}}. \tag{5''}$$

The values of the Q_i are easily obtained from the numerical values of the atomic masses of the different nuclei involved.

We recall that 'atomic mass' is commonly expressed in physics in two different systems of units: the 'physical' system (or the system $\mathrm{O}^{16}=16.000000$) and the system $\mathrm{C}^{12}=12.000000$. In the system $\mathrm{O}^{16}=16$ (the physical system), it is not the atomic mass of the *nucleus* O^{16}, but that of the *atom* O^{16} (with its 8 electrons) that equals 16.000000. Similarly, in the system $\mathrm{C}^{12}=12$, it is the atomic mass of the atom C^{12} with its 6 electrons that equals 12.000000.

Let us place ourselves first in the physical system $\mathrm{O}^{16}=16.000000$. In this system, the unit of atomic mass equals:

$$\mathrm{mA}_{16} = 1.65981 \times 10^{-24}\ \mathrm{g}.$$

According to Einstein's equation, this represents an energy of 1.49177×10^{-3} erg. Expressing this same energy in MeV, we can write (introducing the thousandth part of mA_{16})

$$\boxed{\mathrm{mmA}_{16} = 0.931145\ \mathrm{MeV}}.$$

Then we have for the mass A_e of the electron (β^+ or β^-):

$$A_e = 0.548\,761 \text{ mmA}_{16} = 0.510\,976 \text{ MeV}.$$

Expressing the *nuclear* masses in thousandths of mA_{16} (thousandths of an atomic mass in the 'physical' system) and denoting by ΔA the '*mass excess*'* or the 'fractional' part of the masses of the proton, the neutron, the deuteron, the He^3 nucleus, and the He^4 nucleus, respectively, we have:

$$\begin{aligned}
\Delta A_p &= (A_p - 1) = 7.596\,87 \text{ mmA}_{16} = 7.073\,79 \text{ MeV}\\
\Delta A_n &= (n-1) = 8.976 \text{ mmA}_{16} = 8.358 \text{ MeV}\\
\Delta A_d &= (d-2) = 14.193\,75 \text{ mmA}_{16} = 13.216\,45 \text{ MeV}\\
\Delta A_{He^3} &= (He^3 - 3) = 15.891\,28 \text{ mmA}_{16} = 14.797\,05 \text{ MeV}\\
\Delta A_{He^4} &= (He^4 - 4) = 2.7786 \text{ mmA}_{16} = 2.5873 \text{ MeV}.
\end{aligned}$$

The corresponding values in the system

$$C^{12} = 12.000\,000$$

are obtained by using the following relations:

The unit of atomic mass $\text{mA}_{12} = \text{mA}_{16} \times 1.000\,318$ (in making the conversion, be careful to think in terms of A and not of ΔA).

Since the unit in the C^{12} system is larger, the numerical values of A will be smaller and we will have (with $\text{mmA}_{12} = 0.931\,441$ MeV and A_e equal to the mass of the positive or negative electron); (for *atoms* and not for nuclei!).

$$\begin{aligned}
A_e &= \quad = 0.548\,586 \text{ mmA}_{12} = 0.510\,976 \text{ MeV}\\
\Delta A_p &= (A_p - 1) = 7.276 \text{ mmA}_{12} = 6.777 \text{ MeV}\\
\Delta A_n &= (n-1) = 8.655 \text{ mmA}_{12} = 8.062 \text{ MeV}\\
\Delta A_d &= (d-2) = 14.102 \text{ mmA}_{12} = 13.135 \text{ MeV}\\
\Delta A_{He^3} &= (He^3 - 3) = 16.030 \text{ mmA}_{12} = 14.931 \text{ MeV}\\
\Delta A_{He^4} &= (He^4 - 4) = 2.603 \text{ mmA}_{12} = 2.424 \text{ MeV}.
\end{aligned}$$

The reader should be aware of the fact that while most authors giving tables of atomic masses specify whether they are in the physical system or in the C^{12} system, they generally do not specify whether they are giving the 'atomic weight' of the *atoms* (accompanied by electrons) or the 'atomic weight' of the corresponding *nuclei*. Most tables give (without saying so) the atomic weight of the *atoms*, and the atomic weight of the electrons must be subtracted in order to find the values corresponding to the *nuclei*.

Using the physical system of units ($O^{16} = 16$) and calculating the energy in 'mass excess', we immediately find for the first of the reactions (1):

$$\tfrac{1}{2}Q_1 + \nu = 2\Delta A_p - \Delta A_d - A_e = 0.451\,2 \text{ mmA}_{16} = 0.420\,2 \text{ MeV}.$$

* The purely *mathematical* notion of 'mass excess' (analogous to the fractional part of a logarithm, the only part given in a table) should not be confused with the *physical* idea of 'mass defect' – the difference between the mass of a nucleus and the sum of the masses of its constituent parts. The 'mass defect' represents the energy lost during 'fusion'.

However, the 'neutrino loss' represented by ν is:

$$\nu = 0.282 \text{ mmA}_{16} = 0.263 \text{ MeV}.$$

Thus, we are left with

$$\tfrac{1}{2}Q_1 = 0.169 \text{ mmA}_{16} = 0.157 \text{ MeV}.$$

On the other hand, according to the second of the reactions (1):

$$\tfrac{1}{2}Q_2 = 2A_e = 2 \times 0.5488 \text{ mmA}_{16} = 1.0975 \text{ mmA}_{16} = 1.0221 \text{ MeV}.$$

And finally:

$$\nu + \tfrac{1}{2}Q_1 + \tfrac{1}{2}Q_2 = 1.5487 \text{ mmA}_{16} = 1.4423 \text{ MeV},$$

and

$$\tfrac{1}{2}Q_1 + \tfrac{1}{2}Q_2 = 1.2667 \text{ mmA}_{16} = 1.1793 \text{ MeV}.$$

In the same way, we find, according to (2):

$$\tfrac{1}{2}Q_3 = -\Delta A_{He^3} + \Delta A_d + \Delta A_p = 5.8993 \text{ mmA}_{16} = 5.4933 \text{ MeV},$$

and finally, according to (3):

$$Q_4 = 2\Delta A_{He^3} - 2\Delta A_p - \Delta A_{He^4} = 13.8102 \text{ mmA}_{16} = 12.8593 \text{ MeV}.$$

In sum, we find in this way* that Q_{pp} as defined in (4) is given by:

$$\boxed{Q_{pp} = Q_1 + Q_2 + Q_3 + Q_4 = 28.142 \text{ mmA}_{16} = 26.204 \text{ MeV}}.$$

Note that a single p-p reaction corresponds to only half the chain described above, and consequently releases an energy of only $\tfrac{1}{2}Q_{pp}$.

2.4. The Details of the C–N Cycle (Bethe, 1938)

The cycle begins with the 'radiative capture' of a proton by a nucleus of C^{12}; the energy is released in the form of gamma rays.

$$\boxed{\begin{matrix}\bigcirc\ \bigcirc\ \bigcirc\ \bigcirc\ \bigcirc\ \bigcirc\\ \oplus\ \oplus\ \oplus\ \oplus\ \oplus\ \oplus\end{matrix}} + \oplus \rightarrow \boxed{\begin{matrix}\bigcirc\ \bigcirc\ \bigcirc\ \bigcirc\ \bigcirc\ \bigcirc\\ \oplus\ \oplus\ \oplus\ \oplus\ \oplus\ \oplus\ \oplus\end{matrix}} + Q_1' \qquad (6)$$

$$C^{12} \quad + \text{p} \rightarrow \quad N^{13} \quad + \gamma. \qquad (6')$$

The light isotope N^{13} of the nitrogen nucleus is unstable and disintegrates, giving birth to a positive electron ('β^+ disintegration') and a neutrino (we might say that the 'extra' proton is transformed into a neutron):

* We can also consider that an *atom* of He^4 is produced by the transmutation of 4 *atoms* of hydrogen, with the emission of neutrinos. Then in the (O^{16}) system we have

$$A_{He}(\text{atom}) = 4.003877 \text{ mA}_{16},$$
$$A_H(\text{atom}) = 1.008146 \text{ mA}_{16}.$$

Thus the 'mass defect' equals 0.028707 mA_{16}. Subtracting 2ν, we recover the value of Q_{pp}.

$$\boxed{\begin{array}{l}\bigcirc\bigcirc\bigcirc\bigcirc\bigcirc\bigcirc\\ \oplus\oplus\oplus\oplus\oplus\oplus\oplus\end{array}} \rightarrow \boxed{\begin{array}{l}\bigcirc\bigcirc\bigcirc\bigcirc\bigcirc\bigcirc\bigcirc\\ \oplus\oplus\oplus\oplus\oplus\oplus\end{array}} + \boxed{+} + \boxed{\nu}_1 + Q'_2$$

$$\text{followed by } \boxed{+} + \boxed{-} \rightarrow Q'_3 \qquad (7)$$

$$N^{13} \rightarrow C^{13} + \beta^+ + \nu_1 + Q'_2 \text{ followed by } \beta^+ + \beta^- \rightarrow Q'_3. \qquad (7')$$

(Note that the β^+ disintegration is followed by the annihilation of the β^+ by a negative electron β^-.)

The cycle continues with the fusion of the C^{13} nucleus just formed and another proton, forming a N^{14} nucleus:

$$\boxed{\begin{array}{l}\bigcirc\bigcirc\bigcirc\bigcirc\bigcirc\bigcirc\bigcirc\\ \oplus\oplus\oplus\oplus\oplus\oplus\end{array}} + \oplus \rightarrow \boxed{\begin{array}{l}\bigcirc\bigcirc\bigcirc\bigcirc\bigcirc\bigcirc\bigcirc\\ \oplus\oplus\oplus\oplus\oplus\oplus\oplus\end{array}} + Q'_4 \qquad (8)$$

$$C^{13} \qquad + \text{p} \rightarrow \qquad N^{14} \qquad + \gamma. \qquad (8')$$

The three phases which follow have the same 'structure', except that:

(a) Starting from N^{14} instead of C^{12} introduces an additional $\boxed{\begin{array}{c}\bigcirc\\ \oplus\end{array}}$ group everywhere;

(b) the last reaction (that of O^{16} formation) is accompanied by the 'decomposition' of most of the O^{16} nuclei into C^{12} (which is thus regenerated) and an α particle. (However, a very small number of O^{16} nuclei escape this disintegration, and these nuclei serve as the starting point for a new cycle, called the CNO bi-cycle.)

Thus we have the three following reactions for the C-N cycle, completing the reactions (6), (7), and (8):

(In order to emphasize the analogy between the reactions of the second half of the C-N cycle and those of the first half, we shall give them the same numbers, distinguishing them by an asterisk.)

$$\boxed{© \begin{array}{c}\bigcirc\\ \oplus\end{array}} + \oplus \rightarrow \boxed{© \begin{array}{c}\bigcirc\\ \oplus\oplus\end{array}} + Q'_5 \qquad (6^*)$$

$$N^{14} \quad + \text{p} \rightarrow \quad O^{15} \quad + \gamma. \qquad (6'^*)$$

The light isotope O^{15} of the oxygen nucleus is unstable and disintegrates, giving birth to a positive electron (β^+ disintegration) and a neutrino:

$$\boxed{© \begin{array}{c}\bigcirc\\ \oplus\oplus\end{array}} \rightarrow \boxed{© \begin{array}{c}\bigcirc\bigcirc\\ \oplus\end{array}} + \boxed{+} + \boxed{\nu}_2 + Q'_6$$

$$\text{followed by } \boxed{+} + \boxed{-} \rightarrow Q'_7 \qquad (7^*)$$

$$O^{15} \rightarrow N^{15} + \beta^+ + \nu_2 + Q'_6 \text{ followed by } \beta^+ + \beta^- \rightarrow Q'_7. \qquad (7'^*)$$

Finally, the cycle ends with the fusion of the N^{15} nucleus just formed and another proton, forming a C^{12} nucleus and an α particle (as if the O^{16} that was formed decomposed into C^{12} and He^4):

$$\boxed{\begin{matrix}\bigcirc\,\bigcirc\,\bigcirc\,\bigcirc\,\bigcirc\,\bigcirc\,\bigcirc\,\bigcirc\\ \oplus\,\oplus\,\oplus\,\oplus\,\oplus\,\oplus\,\oplus\end{matrix}} + \oplus \rightarrow \boxed{\begin{matrix}\bigcirc\,\bigcirc\,\bigcirc\,\bigcirc\,\bigcirc\,\bigcirc\\ \oplus\,\oplus\,\oplus\,\oplus\,\oplus\,\oplus\end{matrix}} + \boxed{\begin{matrix}\bigcirc\,\bigcirc\\ \oplus\,\oplus\end{matrix}} + Q'_8 \qquad (8^*)$$

$$N^{15} + p \rightarrow C^{12} + He^4 + Q'_8 . \qquad (8'_*)$$

The values of the Q'_i are* as follows, for the 'neutrino losses' are $\nu_1 = 0.762$ mmA$_{16}$ and $\nu_2 = 1.07$ mmA$_{16}$ respectively (or $\nu_1 = 0.710$ MeV and $\nu_2 = 1.00$ MeV):

$Q'_1 = 2.088$ mmA$_{16}$	$= 1.944$ MeV** and	$= 1.943$ MeV†
$Q'_2 = 0.525$	$= 0.489$	$= 0.489$
$Q'_3 = 1.098$	$= 1.022$	$= 1.022$
$Q'_4 = 8.108$	$= 7.550$	$= 7.549$
$Q'_5 = 7.832$	$= 7.293$	$= 7.291$
$Q'_6 = 0.79$	$= 0.74$	$= 0.74$
$Q'_7 = 1.098$	$= 1.022$	$= 1.022$
$Q'_8 = 5.332$	$= 4.965$	$= 4.964$
$Q' = 26.78$ mmA$_{16}$	$= 25.02$ MeV	$= 25.02$ MeV.

The above values of the Q'_i are easily obtained from the atomic masses of the electron and the proton (in the 'physical' system $O^{16} = 16.000000$), with the addition of the following data for the 'mass excess' of the nuclei of C^{12}, C^{13}, N^{13}, N^{14}, N^{15}, and O^{16}:

$\Delta A(C^{12}) = +0.5224$ mmA$_{16}$	$= +0.4864$ MeV
$\Delta A(C^{13}) = +4.1957$	$= +3.9068$ MeV
$\Delta A(N^{13}) = +6.031$	$= +5.616$ MeV
$\Delta A(N^{14}) = +3.6849$	$= +3.4312$ MeV
$\Delta A(N^{15}) = +1.0356$	$= +0.9643$ MeV
$\Delta A(O^{15}) = +3.450$	$= +3.212$ MeV
$\Delta A(O^{16}) = -4.3901$	$= -4.0876$ MeV.

The mass excess $\Delta A(O^{16})$ of the O^{16} nucleus is *negative*. We also note that in transforming the mass of the atom to that of the nucleus, it is not sufficient (rigourously) to *subtract* the mass of the electrons. One must also *add* the 'binding' energy between the nucleus and the electrons. It can be shown that this energy is of the order of $1.7 \times Z^{7/3} \times 10^{-5}$ mmA$_{16}$ and can, for example, be as great as 0.002 mmA$_{16}$ for $Z = 8$. Thus the fourth decimal place in our values of ΔA is not entirely justified. However, we shall retain it, because more precise nuclear masses can be obtained by the simple addition of the neglected term in $Z^{7/3}$.

We also note that the 'result' of the C-N cycle, like that of the p-p chain (represented by Equation (5')) does not modify the electrical neutrality of the plasma made up of the protons and negative electrons arising from the ionization of the hydrogen atoms.

* By the argument used for the 'direct' calculation of Q_{pp}, we recover the value of Q' by subtracting $(\nu_1 + \nu_2) = 1.832$ mmA$_{16}$ from the 'mass defect' of 28.707 mmA$_{16}$.

** According to [2], p. 542.

† According to [4], Table 6.

For of the 4 electrons liberated by the ionization of 4 hydrogen atoms (while the 4 protons go into the fusion of a helium nucleus (He^4)), two electrons are annihilated by the β^+ and the other two are neutralized by the charge $Z=2$ of the He^4 nucleus that is formed. Electrically, each group of 4 hydrogen atoms originally represents a plasma containing 4 elementary positive charges (the 4 protons) and 4 elementary negative charges (the 4 negative electrons). The final result is to transform this plasma into a He^4 nucleus (two elementary positive charges) and two negative electrons (two elementary negative charges) – neglecting the γ-rays and the neutrinos, which have no charge.

3. Calculation of the Energy ε. Generalities

The thermonuclear reactions to be studied are chains or cyclic reactions in a steady state.

The determination of the energy ε produced per second by each gram of the stellar material consists of the following steps:

(1) The calculation of the average number of reactions per cm^3 per second for each part of a particular chain or cycle. The theoretical or experimental determination of certain constants appearing in this calculation.

(2) The theoretical or experimental determination of the quantity of energy released by each reaction in the chain or cycle. Addition of the energies released by all the reactions of *one* chain or *one* cycle, taking into account certain losses (neutrino losses).This is the calculation made in Section 2.

(3) Analysis of the adjustment of the various reactions of a given cycle to a stationary state, and calculation of the 'equilibrium abundances' of the various constituents of the mixture.

So let us first consider any one of the reactions of a chain or a cycle, and try to determine the number of reactions R_{12} per cm^3 per second. (R_{12} is called the 'reaction rate' per cm^3 per second.)

3.1. Calculation of R_{12} for a given reaction

We assume that the following data are known:

(1) The nature of the reacting particles. Each of the reactions considered is at most a 'binary' reaction between two constituents. Let (1) and (2) be the indices characterizing each of them. The 'nature' of the nuclei (1) and (2) is characterized by their *atomic numbers* (electric charge in units of the charge of the proton) Z_1 and Z_2, and by their *atomic weights* (more correctly, atomic *masses*) A_1 and A_2. The A_i represent the ratio of the mass of each nucleus (i) to the unit of atomic mass (in the system $O^{16}=16$ for example).

(2) The *densities* in numbers of particles per cm^3 n_1 and n_2 – that is, the composition of the mixture in terms of the reacting particles (1) and (2).

These data make it possible to calculate the theoretical expression for the coefficient σ_{12} (which for brevity we shall denote by σ alone; this will not introduce any uncertainty as long as we limit ourselves to two well-determined constituents), appearing in the expression*

$$R_{12} = \sigma n_1 n_2 V = (n_1 V)\,\Sigma_2 \tag{9}$$

for the number of reactions per cm^3 per second as a function of n_1, n_2, and the relative velocity V (more exactly, the *modulus* of the relative velocity) of the particles. We see that σ represents a generalization of the idea of (microscopic) 'collision cross-section'.

The coefficient σ appears in Equation (9) because not all of the 'encounters' between a particle of type (1) and a particle of type (2) necessarily lead to a reaction of the type under consideration between the two particles. In fact, at the time of such an encounter only a fraction of the particles in collision succeeds in overcoming the electrostatic *repulsion* between the two positively charged nuclei. This fraction is determined by a 'quantum' effect (tunnel effect) – the quantum mechanical wave function ψ (the square of whose modulus represents the probability 'density' of the presence of the particle) is not subject to the 'potential barrier' on which the repulsion depends in as restrictive a manner as the particles associated with this wave. It is shown in quantum mechanics that this fraction is given by the following formula, established for the first time by G. Gamow (widely known for his popular science books):

$$f_G(W) = \exp\left[-CW^{-1/2}\right]. \tag{10}$$

C is a constant given by

$$C = 31.3\, Z_1 Z_2 \left[A_1 A_2/(A_1 + A_2)\right]^{1/2}, \tag{11}$$

taking the reference system connected to the 'center of mass', with

$$W = \tfrac{1}{2}MV^2 \quad \text{in } \textit{kiloelectron-volts}, \tag{12}$$

where

$$M = m_{\mathrm{H}} A_1 A_2/(A_1 + A_2). \tag{13}$$

Remark. Some authors (for example, Fowler) write Equation (10) in the form

$$\exp\left[-(E_G/E)^{1/2}\right]$$

and give the 'Gamow energy' E_G, which is identical to C^2.

These formulae show that the 'Gamow fraction' $f_G(W)$ increases when $Z_1 Z_2$ decreases; it is larger for nuclei of small charge (weak repulsion) which are also *light nuclei.* For a reaction between two protons

$$(Z_1 = Z_2 = A_1 = A_2 = 1)$$

we have

$$C_{\mathrm{pp}} = 22.2, \tag{14}$$

* For a better understanding of Equation (9) see [9], Chap. III, Equation (12); replace p_{tot} by R_{12}, Φ by $n_1 V$, and Σ by σn_2.

whereas for a reaction between a carbon nucleus C^{12} (of charge $Z_1=6$) and a proton, we have

$$C_{pc} = 22.2 \times 8.1 \quad [\text{for } (12/13)^{1/2} = 0.96]. \tag{15}$$

The Gamow fraction $f_G(W)$ also increases when the 'relative' kinetic energy W increases; for $W=20$ keV and for $W=2000$ keV$=2$ MeV respectively, the 'proton-proton' reaction gives

$$f_G = \exp(-5.0) = 0.007 \quad \text{and} \quad f_G = \exp(-0.5) = 0.60.$$

Nevertheless, among the nuclei which succeed in overcoming the electrostatic repulsion, only a fraction really 'reacts'; this new fraction also depends on the 'nature' of the reacting nuclei and on W. But here, in contradiction to what happens for the Gamow fraction, the fraction decreases when W increases. (This property recalls the well-known efficiency of slow neutrons in fission reactions. More generally, when the relative velocity of the particles decreases, the probability of an interaction between them *increases*, because they remain 'in contact' longer.)

This new fraction is of the form $S(W)/W$, where $S(W)$ is, for values of W far from certain critical values called 'resonances', a function which varies very slowly with W. Fowler represents it by a Taylor series:

$$S(W) = S(0)\left[1 + \frac{S'(0)}{S(0)}W + \frac{1}{2}\frac{S''(0)}{S(0)}W^2\right],$$

which for usual values of W reduces to $S(0)=S_0$.

Thus σ is given by the formula:

$$\sigma = (S_0/W)\, f_G(W), \tag{16}$$

when we reduce the function $S(W)$ to $S(0)=S_0$.

The quantity S_0 is a numerical constant which depends on the nature of the nuclei in the reaction, and on the *type of reaction* between these nuclei. It is generally determined by laboratory experiments. But an exception is made for the reaction between protons, where it is calculated theoretically; this is possible because the case of protons is particularly simple, and necessary because S_0 is particularly small in this case, and the interaction effect is difficult to observe in the laboratory when S_0 is very small. (Fowler, *loc. cit.*, gives the coefficients for the transformation from $S(0)$ to $S(W)$, when this is necessary.)

Equation (16) is not applicable in the neighbourhood of certain values of W called 'resonances', which are different for each type of reaction and for which σ is much greater than the value given by (16). S_0 is expressed in units called 'keV $\times$ barn', equivalent to

$$1.6 \times 10^{-33} \text{ ergs} \times \text{cm}^2.$$

When n_1 and n_2 are measured in number of particles per cm^3 and V in cm/sec, R_{12}

(the number of reactions per cm^3 per second) is obtained by computing from Equation (16) with W in keV and S_0 in keV $\times$ cm^2, making use of the definition of the barn:

$$\boxed{\text{one barn} = 10^{-24}\ \text{cm}^2}\,.$$

The values of S_0 given by Fowler ([2], Table III) are presented in Table XII.*

TABLE XII

(Typical) reaction	S_0 (keV $\times$ barn)
$H^1(p, \beta^+, \gamma)\ D^2$	3.36×10^{-22} [a]
$D^2(p, \gamma)\ He^3$	2.5×10^{-4}
$He^3(He^3, 2p)\ He^4$	$5.0 \times 10^{+3}$
$C^{12}(p, \gamma)\ N^{13}$	1.40
$C^{13}(p, \gamma)\ N^{14}$	5.50
$N^{14}(p, \gamma)\ O^{15}$	2.75
$N^{15}(p, \alpha)\ C^{12}$	5.34×10^4

[a] According to *Astrophys. J.* **155** (1969), 507, $S_0 = (3.78 \pm 0.15) \times 10^{-22}$ for this reaction.

A symbol such as $C^{12}(p, \gamma)N^{13}$ represents a reaction in which a nucleus of C^{12} reacts with a proton p to form the nitrogen isotope N^{13}, with the emission of a photon γ. The symbol β^+ represents a positive electron.

As we have just seen, the number of reactions R_{12} of a given type depends on the relative velocity of the reacting nuclei, among other factors. Now at a given temperature T, *all* relative velocities V, distributed according to a certain law, are to be found in a mixture of particles. The distribution of relative velocities is easily calculated from the velocity distribution (assumed to be Maxwellian) of particles of type (1) and (2) respectively.

If the number of particles (1) with a velocity modulus V_1 between V_1 and $(V_1 + \mathrm{d}V_1)$ is denoted by $\mathrm{d}n_1$, we have

$$\mathrm{d}n_1 = n_1 f(V_1, M_1, T) \cdot \mathrm{d}V_1 \,. \tag{17}$$

And similarly,

$$\mathrm{d}n_2 = n_2 f(V_2, M_2, T) \cdot \mathrm{d}V_2 \,, \tag{18}$$

where M_1 and M_2 are the masses in grams of particles of type (1) and (2), respectively. The function

$$f(V, M, T) = \left[\frac{M}{2\pi kT}\right]^{3/2} 4\pi V^2 \exp\left[-\frac{1}{2}\frac{MV^2}{kT}\right] \tag{18'}$$

is also found in the expression for the number $\mathrm{d}N_{12}$ of particles of types (1) and (2)

* The use of these values of S_0 with W in keV gives σ in barns. To obtain σ in cm^2, the values of S_0 given in the table should be multiplied by 10^{-24}.

having a relative velocity modulus between V and $(V+\mathrm{d}V)$; but M is replaced by the 'reduced' mass

$$M = M_1 M_2/(M_1 + M_2), \tag{19}$$

which has already been given in terms of A_1 and A_2 in Equation (13).

Thus we have

$$\mathrm{d}N_{12} = n_1 n_2 f(V, M, T)\,\mathrm{d}V, \tag{20}$$

which replaces the factor $n_1 n_2$ in Equation (9) when R_{12} is replaced by $\mathrm{d}R_{12}$, the number of reactions per cm^3 per second corresponding to the interval $(\mathrm{d}V)$ in the neighbourhood of V:

$$\mathrm{d}R_{12} = \sigma n_1 n_2 f(V, M, T)\, V\,\mathrm{d}V. \tag{21}$$

With W as the independent variable instead of V, we have:

$$W = \tfrac{1}{2}MV^2 \quad \text{and} \quad W\,\mathrm{d}W = \tfrac{1}{2}M^2V^3\,\mathrm{d}V. \tag{22}$$

Now, the replacement of $f(V, M, T)$ by the expression given above introduces the product $V^3\mathrm{d}V$ – that is, $W\mathrm{d}W$ – in (21). But the replacement of σ by expression (16) introduces W in the denominator, and there remains only the factor $(\mathrm{d}W)$. Thus we find:

$$\mathrm{d}R_{12} = \frac{S_0 n_1 n_2 C_0\, \mathrm{e}^{-W/kT} f_G}{(kT)^{3/2}}\,\mathrm{d}W = \frac{S_0 n_1 n_2 C_0}{(kT)^{3/2}} \exp\left[-\frac{W}{kT} - CW^{-1/2}\right]\mathrm{d}W, \tag{23}$$

where C_0 represents all the factors depending on M.

The integration of $\mathrm{d}R_{12}$ with respect to W between $W=0$ and $W=\infty$ is a purely mathematical problem, facilitated by the form of the function to be integrated. Indeed, the *exponent* of the exponential in (23) is a function which passes through a very sharp maximum for a certain value of W. In the neighbourhood of this value of W we can expand this function in a Taylor series and, retaining only the first three terms, use this limited expansion as an approximate representation of the real function (the curve representing the real function is thus replaced by a parabola in the neighbourhood of the maximum). Far from the maximum in question the approximate representation is incorrect, but it happens that for usual values of the temperature the contribution of these parts of the function to the integral appearing in R_{12} is negligible in any case, both for the exact function and for its 'parabolic' representation.

Let us therefore put

$$W/kT = \alpha t, \tag{24}$$

where α is a constant which we shall use to put the function $W/kT + CW^{-1/2}$ in the form

$$\frac{W}{kT} + CW^{-1/2} = \alpha\,(t + t^{-1/2}). \tag{25}$$

It is readily seen that in order to obtain this identity we must take

$$\alpha = C^{2/3}(kT)^{-1/3}. \tag{26}$$

We also see that the function $(t+t^{-1/2})$ passes through a minimum (which corresponds to a maximum of the same function with a change in sign) for $t=t_0=0.63$. The expansion in the neighbourhood of $t=t_0$ can be put in the approximate form:

$$t + t^{-1/2} = 1.89 + 1.17(t - 0.63)^2. \tag{27}$$

Using the well-known integral $\int_0^\infty \exp(-t^2)\, \mathrm{d}t$ and neglecting a small contribution to the integral between 0 and t_0, we find the following expression for ε_{12} in ergs per gram per second:

$$\varepsilon_{12} = Q'R_{12}/\varrho = CQ\varrho X_1 X_2 S_0 D T_6^{-2/3}\, \mathrm{e}^{-B}, \tag{28}$$

where Q is the energy in MeV and Q' the energy in ergs, produced by each elementary reaction; $C=7.567\times 10^{26}$; ϱ is the density of the mixture; X_i is the abundance $m_H A_i \varrho^{-1} n_i$ in grams of (i) *per gram of mixture*; $D=Z_1^{1/3}Z_2^{1/3}/A_1A_2A^{1/3}$; $A=M/m_H$; $B=42.48\ [Z_1^2Z_2^2A/T_6]^{1/3}$; T_6 is the temperature in millions of degrees Kelvin and S_0 is in keV $\times$ barn. (*Remark*: Fowler [2] writes our B as τ and our $Z_1^2Z_2^2A$ as W).

4. The 'Mean Lifetime' of a Given Nucleus with Respect to an Isolated Reaction (R)

4.1. Generalities

To be specific, let us consider the reaction (R) of the 'beginning' of the C-N cycle: $C^{12}+p\rightarrow N^{13}+\gamma$.

Let n_c be the (number) density of C^{12} nuclei and n_p the (number) density of protons. We recall that $A_c=12.000552$, $Z_c=6$, $A_p=1.007597$ and $Z_p=1$. Here we denote by $M_U=mA_{16}=1.65981\times 10^{-24}$ grams the unit of 'atomic' (and nuclear) mass on the scale (complete *atom* of $O^{16}=16$). Then A_p will be the mass of a hydrogen *nucleus* (proton) in units of M_U, and A_c will be the mass of a C^{12} *nucleus* in units of M_U.

Since each reaction (R) has the effect (among others) of *decreasing* the numbers n_c and n_p, the densities n_c and n_p will vary with time. Let $n_c(t)$ and $n_p(t)$ be the functions describing these variations.

Let R_{cp} be the number of reactions (R) per cm^3 per second. (More generally, the subscripts c and p can also be interpreted as referring to the 'target' and the 'projectile', respectively.) If the target is given the subscript (0) and the projectile the subscript (1), the number of reactions R_{cp} will be written as R_{01}. (*Remark*. Fowler [2] refers to our R_{01} as P_{01}. In this case, P represents the first letter of the word 'process'.)

In addition, let $(\mathrm{d}n_c)_p$ be the (negative) variation of n_c for reactions with the particles p during the interval $(\mathrm{d}t)$, and let $(\mathrm{d}n_p)_c$ be the (negative) variation of n_p for reactions with the particles c.

We emphasize, once again, that we are neglecting for the moment any other possible causes for the variation of n_c and n_p. Only the effects of the reactions (R) are considered.

We obviously have:

$$(\mathrm{d}n_c)_p = -R_{cp}\,\mathrm{d}t\,. \tag{29}$$

But we must not forget that R_{cp} depends in turn on n_c, through the relation

$$R_{cp} = S_0 F_{cp} n_c n_p\,, \tag{30}$$

where F_{cp} depends on T, A_c, A_p, Z_c, and Z_p, as shown by Equation (28). Of course, S_0 has the value characteristic of the reaction (R).

To obtain a relation of the same form as Equation (9), we set (after Fowler):

$$S_0 F_{cp} = \langle \sigma V \rangle_{cp} = \langle cp \rangle\,. \tag{31}$$

Equation (30) can then be written

$$R_{cp} = \langle cp \rangle\, n_c n_p\,, \tag{32}$$

or, dividing the two members of (29) by n_c,

$$[\mathrm{d}n_c/n_c]_p = -\langle cp \rangle\, n_p\,\mathrm{d}t\,. \tag{33}$$

The symbol $[\cdots]_p$ indicates the nature of the projectiles.

In general, under stellar conditions the 'reservoir' of protons is so great that the variation of n_p can be neglected over a long interval Δt. Moreover, the 'heat capacity' of the stellar material is such that the variation ΔT of the temperature remains negligible during Δt, and T can be considered constant. This is true for intervals Δt of the order of several million years.

For longer intervals of time, one can generally think of the evolution as a succession of states of the same type, but with different values of n_p and T for each stage.

Thus, the ratio n_c/R_{cp} is a constant over Δt, and is equal to

$$n_c/R_{cp} = 1/S_0 F_{cp} n_p\,. \tag{34}$$

Putting

$$n_c/R_{cp} = \tau_p(c) = 1/\langle cp \rangle\, n_p\,, \tag{35}$$

we can write (33) in the form:

$$[\mathrm{d}n_c/n_c]_p = -\,\mathrm{d}t/\tau_p(c)\,. \tag{36}$$

Since the left-hand side of (36) is obviously 'dimensionless', we see that $\tau_p(c)$ has the dimensions of an interval of *time*.

Integrating the differential Equation (36) with respect to the time t and letting n_c^0 be the value $n_c(0)$ of n_c at the time $t=0$, we obtain

$$n_c(t) = n_c^0\, \mathrm{e}^{-t/\tau_p(c)}\,. \tag{37}$$

Under stellar conditions, $\tau_p(c)$ is of the order of 10^6 to 10^9 years.

We emphasize once again that $n_c(t)$ as given by (37) corresponds only to a 'partial' rate of variation – the one resulting from the reaction (R) to the exclusion of all other reactions. In a more complete and thorough study, one must, of course, take into account all the reactions capable of producing a variation of n_c (and, especially for high temperatures, the 'inverse' reactions which limit – in nuclear physics as in organic chemistry – the 'yield' of each 'direct' reaction. Fowler's review paper [2], which we have already cited many times, contains all the necessary information to correct for inverse reactions.)

The physical meaning of the parameter $\tau_p(c)$ is generally considered 'obvious', but this is not at all the case. We shall devote the following section to an examination of this meaning.

4.2. The physical meaning of $\tau_p(c)$

We can imagine each disappearance of $(\mathrm{d}n_c)_p$ 'target nuclei' (here C^{12}) as the 'death' of these nuclei, a death ascribable to the projectiles p. Let us consider an interval of time $\mathrm{d}t$ equal to *one year*, which represents a *very short* interval for the variation of $(\mathrm{d}n_c)_p$. In fact, using Fowler's data ([2], p. 554) we find $\tau_p(c)=1.7\times 10^7$ years (for $T=13\times 10^6$ K and $\varrho X_p=100$). Thus, under the prevailing conditions near the center of the Sun, the fraction of C^{12} nuclei transformed into N^{13} nuclei in one year under the influence of the projectiles p is of the order of $\mathrm{d}t/\tau_p(c)=6\times 10^{-8}$.

The $|\mathrm{d}n_c|_p$ carbon nuclei which 'die' in the interval of time $(t, \mathrm{d}t)$ *are exactly those which*, to second-order accuracy, *have 'lived' t seconds from the time* $t=0$ – *those whose 'lifetime' was t seconds.*

Now, in a family, if child A dies at the age of 13, child B at the age of 24, the father at the age of 67, and the mother at the age of 56, the mean of the lifetimes of a member of this family would be $(13+24+56+67)/4=40$ years. As we see in this example, the mean of the lifetimes of a member of a given family is obtained by dividing the sum of the lifetimes of all the members of the family by the number of members.

But for our carbon nuclei, the *sum* of the lifetimes of those that 'die' between the time t and the time $(t+\mathrm{d}t)$ is the *product* $t\cdot|(\mathrm{d}n_c)_p|$ of the absolute value of $(\mathrm{d}n_c)_p$ and t. That is, the product of the lifetime t of each of them and the number of disappearances taking place in $(t, \mathrm{d}t)$. This sum is thus a function of t. Moreover, since $(\mathrm{d}n_c)_p$ is negative, we have

$$|(\mathrm{d}n_c)_p| = -(\mathrm{d}n_c)_p .$$

Adding up all these sums, we obtain an essentially positive quantity

$$-\int_0^\infty t\cdot(\mathrm{d}n_c)_p ,$$

which represents the overall sum of all the lifetimes of the carbon nuclei. Dividing by n_c^0, we have the mean of the lifetimes – or, as we say somewhat inexactly, 'the mean lifetime' – of the carbon nuclei which have been exposed from the time $t=0$ to bombardment by the p.

This 'mean time' will thus be given, according to its *physical* definition, by the equation:

$$\text{'mean lifetime of a nucleus } (c)\text{'} = \frac{1}{n_c^0} \int_0^\infty - t \cdot (\mathrm{d}n_c)_p . \tag{38}$$

But we obtain from Equation (37), using the differential relation (36) to simplify the calculation,

$$- (\mathrm{d}n_c)_p = \frac{n_c(t) \cdot \mathrm{d}t}{\tau_p(c)} = \frac{n_c^0 \, \mathrm{e}^{-t/\tau_p(c)} \, \mathrm{d}t}{\tau_p(c)} ; \tag{39}$$

whence according to (38) we have the following expression for the 'mean lifetime':

$$\frac{1}{\tau_p(c)} \int_0^\infty t \cdot \mathrm{e}^{-t/\tau_p(c)} \, \mathrm{d}t . \tag{40}$$

Or again, putting $x = t/\tau_p(c)$:

'Mean lifetime of (c) with respect to the isolated reaction (R)' =

$$\tau_p(c) \int_0^\infty x \, \mathrm{e}^{-x} \, \mathrm{d}x = \tau_p(c) \cdot (1!) = \tau_p(c) . \tag{41}$$

Thus we find that the parameter $\tau_p(c)$ of Equation (36) is nothing but the mean of the lifetimes – or the 'mean lifetime' – of the carbon nuclei of the stellar mixture with respect to reaction (R), produced by a bombardment with protons as projectiles. This is the mean lifetime which the C^{12} carbon nuclei *would have had* if interactions of the type $C^{12}(p, \gamma)\, N^{13}$ had been the only ones modifying n_c. But in reality, the C^{12} nuclei are 'regenerated' by reactions other than (R).

In addition, since $\tau_p(c)$ changes in the interval of time from $t=0$ to $t=\infty$, the preceding interpretation is *purely conventional.* To be rigourous, we must say, "$\tau_p(c)$ *would be* the mean lifetime of the carbon nuclei if $R_{cp}/n_c = S_0 F_{cp} n_p$ remained constant".

Because of continuing work on the experimental values of these quantities, S_0 'changes' with time, and consequently the value of $\tau_p(c)$ changes with each new publication. Thus, for $T_6=13$ and $\varrho X_p=150$, Schwarzschild [1], (1958), p. 76, gives $\tau_p(C^{12}) = 41 \times 10^{13}$ seconds. But according to Reeves [4], (1965), p. 144, this parameter equals 37×10^{13} seconds. And according to the more recent work of Fowler [2], (1967), we have $\tau_p(C^{12}) = 32 \times 10^{13}$ seconds. It is a real 'magic skin'!

4.3. $\tau_p(c)$ AS THE 'MEAN DURATION OF AN ISOLATED REACTION (R)'

From another point of view, the 'lifetime' t of the $\mathrm{d}n_c(t)$ particles (c) destroyed between the time t and the time $(t+\mathrm{d}t)$ by the reaction (R) can also physically represent the *waiting* time before participation in the reaction (R). This is why it is sometimes said that $\tau_p(c)$ represents the 'mean duration of the reaction R'. But this expression may wrongly suggest that we are dealing with the mean value of the *durations* (dt) *of the reactions themselves.*

4.4. $\tau_p(c)$ AS AN 'EXPONENTIAL DECREMENT'

We note that – independently of its meaning as a 'mean lifetime' – $\tau_p(c)$ also clearly represents (according to (37)) the time that would be required for the density $n_c(t)$ to decrease exponentially to approximately one-third of its value at $t=0$, if the density depended only on interactions between the particles (c) and the particles (p) (and if, in addition, the parameter $\tau_p(c)$ itself were constant.) For we have $e^{-1}=1/2.7\ldots$.

4.5. THE TRANSITION PROBABILITY p_{ca} PER REACTION (R)

Let us assume that the reactions (R) transform the 'target' particles (c) into particles (a) (in the present case, into nitrogen nuclei N^{13}). More generally, we can also say that the reactions (R) change the particles (c) to a new 'state' (a).

Now Equation (36) can be written, putting

$$p_{ca} = 1/\tau_p(c), \tag{42}$$

in the more workable form:

$$(dn_c)_p = - p_{ca} n_c \, dt. \tag{43}$$

(*Remark*. The authors of reference [8] denote our particles (c) by (0), and our projectiles (p) by (1), writing p_{ca} in the form $p_1(0)$. Fowler [2], in order to avoid confusion with his P_{01} which represents our R_{cp}, is obliged to introduce a particularly unfortunate notation for our p_{ca}: $\lambda_1(0)$. Reeves denotes our p_{ca} by P_{cp}.)

Writing (43) in the form

$$p_{ca} = [(- dn_c)_p / n_c] / dt, \tag{44}$$

we see that p_{ca} represents the *fraction* of the n_c transformed into (a) per second. This quantity can therefore be interpreted as the probability of transforming *one* particle (c) into a particle (a), or as a '*transition probability*' $(c) \rightarrow (a)$ which is measured in s^{-1}. We have:

$$p_{ca} = R_{cp}/n_c. \tag{45}$$

This also enables us to interpret p_{ca} as '*the mean number of reactions per nucleus* (c) *per second*' (or '*mean interaction rate per nucleus of type* (c)'). Similarly, we say that $\tau_p(c)$ represents the '*mean reaction time per nucleus of type* (c)'.

5. The Convergence of Cyclic Reactions to a Stationary ('Equilibrium') State

To obtain a better understanding of what happens in a *chain of cyclic reactions* (such as the C-N cycle), we can imagine the simple theoretical case of *two* components (c) and (a) which are cyclically connected.

Let the particles (c), for example, be nuclei of C^{12}, and let the particles (a) be nuclei of N^{14}; we assume that they can react according to the following scheme:

Reaction (R): $(c)\rightarrow(a)$.
Reaction (R′): $(a)\rightarrow(c)$.

In the case of the C-N cycle, reaction (R) represents the ensemble of the three 'initial' reactions (6), (7) and (8) of the cycle, while (R′) represents the three 'final' reactions (6*), (7*), and (8*); the latter set 'regenerates' C^{12} from the N^{14} formed by the three 'initial' reactions.

Of course, the present study can also be applied to the convergence towards equilibrium of two '*states*' (c) and (a) of a single type of atom (neutral or ionized), under the influence of the absorption and emission of photons – e.g. between the 'fundamental' state and the first excited state (resonance transitions).

For maximum conciseness, we denote by c the number n_c of particles (c) per cm^3 at the time t; and likewise we denote by a the number n_a of particles (a) per cm^3 at the time t. The problem is to find the functions $c(t)$ and $a(t)$ which describe the variation of c and a with the time t.

Now, during the interval dt which follows the time t, the increase in c due to reactions $(a)\rightarrow(c)$ is equal to the decrease $|dn_a|$ and can be expressed as a *positive* quantity $(-dn_a)$†; similarly, the decrease in c due to reactions $(c)\rightarrow(a)$ is given by the *positive* quantity $(-dn_c)$. In other words, the total algebraic variation (positive or negative) of c during dt is given by:

$dc=$[increase due to reactions $(a)\rightarrow(c)$]$-$[decrease due to reactions $(c)\rightarrow(a)$].

Thus, dropping the subscript p in order to simplify the notation,

$$dc = (-\,dn_a) - (-\,dn_c). \tag{46}$$

However, according to (43) we have

$$(-\,dn_a) = n_a p_{ac}\,dt = a p_{ac}\,dt, \tag{47}$$

and

$$(-\,dn_c) = n_c p_{ca}\,dt = c p_{ca}\,dt. \tag{48}$$

To be even more concise (without any risk of ambiguity), we put

$$p_{ac} = p_a \quad \text{and} \quad p_{ca} = p_c. \tag{49}$$

We then have for (dc) the expression

$$dc = (a p_a - c p_c)\,dt, \tag{50}$$

which can be put in the form

† We use here (dn_a) as before for a description of 'partial' reactions (R′).

$$\mathrm{d}c/\mathrm{d}t = ap_a - cp_c\,. \tag{51}$$

We could, of course, establish an analogous relation for $\mathrm{d}a/\mathrm{d}t$ by a similar line of reasoning. We would thus obtain two differential equations for the two unknown functions $a(t)$ and $c(t)$. But taking into account the cyclic nature of the reactions, and assuming that each particle (a) yields only one particle (c) and vice versa, we have an obvious 'first integral' which expresses the conservation of the total number of particles and which is written (denoting the 'initial' densities by $a_0 = a(0)$ and by $c_0 = c(0)$ and their sum by s_0):

$$a(t) + c(t) = a_0 + c_0 = s_0 = \text{const}\,. \tag{52}$$

Thus at all times t we have

$$a(t) = s_0 - c(t)\,, \tag{53}$$

and the problem reduces to the determination of $c(t)$ by the single differential equation

$$\mathrm{d}c/\mathrm{d}t = (s_0 - c)\,p_a - cp_c\,. \tag{54}$$

Once $c(t)$ is obtained by the integration of this equation, $a(t)$ will be given by (53).

A new parameter of interest can be introduced by writing (54) in the form

$$\mathrm{d}c/\mathrm{d}t = s_0 p_a - (p_a + p_c)\,c\,, \tag{55}$$

which leads us to define the new parameter p by*:

$$p = p_a + p_c = p_{ac} + p_{ca}\,. \tag{56}$$

Thus we have a linear first-order differential equation with constant coefficients

$$\mathrm{d}c/\mathrm{d}t + pc = s_0 p_a\,, \tag{57}$$

whose classical solution immediately yields (letting $c_\infty = c(\infty)$ be the value of c when $t \to \infty$)

$$c(t) = c_0\,\mathrm{e}^{-pt} + c_\infty\,(1 - \mathrm{e}^{-pt})\,. \tag{58}$$

The value of c_∞ is obtained in terms of the initial conditions $s_0 = a_0 + c_0$, and of the constant known parameters p and p_a, by requiring (58) to satisfy the differential Equation (57) identically. But from (58) we have

$$\begin{aligned} \mathrm{d}c/\mathrm{d}t &= -\,c_0 p\,\mathrm{e}^{-pt} + c_\infty p\,\mathrm{e}^{-pt}\,, \\ pc &= +\,c_0 p\,\mathrm{e}^{-pt} + c_\infty p\,(1 - \mathrm{e}^{-pt})\,. \end{aligned} \tag{59}$$

* No confusion with p = proton (or p = projectile) is possible here.

The substitution of these expressions in the left-hand side of (57) gives, by identity with the right-hand side:

$$c_\infty p = s_0 p_a. \tag{60}$$

Whence

$$c_\infty = s_0 p_a / p. \tag{61}$$

In order to shorten the notation as much as possible, let us introduce the functions

$$D(t) = \mathrm{e}^{-pt} \tag{62}$$

and

$$S(t) = 1 - \mathrm{e}^{-pt} = 1 - D(t). \tag{63}$$

The first function describes an exponential *decrease* (of decrement $1/p$) from 1 to 0, while the second describes a 'complementary' exponential increase towards *saturation*: $S(\infty) \to 1$. It is clear that at any time t, we have

$$D(t) + S(t) = 1. \tag{64}$$

Thus we can put (58) in the elegant form

$$c(t) = c_0 D(t) + c_\infty S(t), \tag{65}$$

which clearly shows how $c(t)$ varies, *following a relatively complicated law*, from c_0 to c_∞.

By symmetry, we can predict that $a(t)$ varies according to the law

$$a(t) = a_0 D(t) + a_\infty S(t). \tag{66}$$

But it is easy to prove this, and at the same time to obtain an expression for a_∞. We have merely to add (65) and (66), obtaining

$$a(t) + c(t) = (a_0 + c_0)\, D + (a_\infty + c_\infty)\, S = s_0 D + (a_\infty + c_\infty)\, S. \tag{67}$$

But according to (52) the sum $(a+c)$ must always remain equal to s_0; therefore we must have

$$a_\infty + c_\infty = s_0. \tag{68}$$

It is easy to show that this relation transforms (67) into an identity, since $D+S=1$. Thus we have

$$a_\infty = s_0 - c_\infty = s_0 - s_0 p_a/p = s_0 (p - p_a)/p = s_0 p_c / p. \tag{69}$$

In summary:

$$c_\infty = s_0 p_a / p \quad \text{and} \quad a_\infty = s_0 p_c / p. \tag{70}$$

It is now advantageous to return to the parameters $\tau_p(c)$ and $\tau_p(a)$, suppressing the subscript to simplify the notation. From (42) we have

$$p_c = p_{ca} = 1/\tau(c) \quad \text{and} \quad p_a = p_{ac} = 1/\tau(a). \tag{71}$$

And using (56), we can write p as

$$p = \frac{\tau(a) + \tau(c)}{\tau(a)\,\tau(c)}. \tag{72}$$

This leads us to introduce a new parameter τ_{cycle}, defined by

$$\tau_{\text{cycle}} = \tau(a) + \tau(c). \tag{73}$$

Then Equations (70) for c_∞ and a_∞ take on the form

$$c_\infty = s_0 \tau(c)/\tau_{\text{cycle}}, \tag{74}$$

and

$$a_\infty = s_0 \tau(a)/\tau_{\text{cycle}}. \tag{75}$$

In other words, we have

$$a_\infty/\tau(a) = c_\infty/\tau(c) = s_0/\tau_{\text{cycle}}, \tag{76}$$

or

$$a_\infty/c_\infty = \tau(a)/\tau(c), \tag{77}$$

with

$$\begin{aligned} a_\infty &= (a_0 + c_0)\,\tau(a)/\tau_{\text{cycle}} \\ c_\infty &= (a_0 + c_0)\,\tau(c)/\tau_{\text{cycle}}. \end{aligned} \tag{78}$$

Thus there is always *convergence* towards a stationary state, reached when $t \to \infty$, in which the densities a_∞ and c_∞ are *proportional* to the mean lifetimes of the corresponding particles with respect to the reactions (R) and (R′), when these are assumed to take place in isolation. The 'hardiest' species of particules – the one that is slowest to disappear – dominates after an infinite time. The same remark applies, in atomic physics, to the predominance of the most 'stable' state.

If we now return to the fundamental Equation (51), introducing the solutions (65) and (66), we easily find (after some elementary calculations, in which we use the values of a_∞ and c_∞ given by (70) and denote $\tau(a)$ by τ_a and $\tau(c)$ by τ_c):

$$\frac{\mathrm{d}c}{\mathrm{d}t} = \left[\frac{a_0}{\tau_a} - \frac{c_0}{\tau_c}\right] D(t) = \left[\frac{a_0}{\tau_a} - \frac{c_0}{\tau_c}\right] \mathrm{e}^{-pt}. \tag{79}$$

This equation gives the law of variation of $\mathrm{d}c/\mathrm{d}t$.

It shows that if we start with a distribution (a_0, c_0) identical with that which would occur automatically (but after an infinite time) – that is, a distribution which satisfies the equivalent conditions

$$a_0/\tau_a = c_0/\tau_c \tag{80}$$

or

$$a_0 = a_\infty \quad \text{and} \quad c_0 = c_\infty \tag{81}$$

– we have identically, starting with the time $t=0$ and for any time t,

$$\mathrm{d}c/\mathrm{d}t = 0 \quad \text{and} \quad \mathrm{d}a/\mathrm{d}t = 0\,. \tag{82}$$

Thus the distribution remains *stationary*. We express this fact by saying that the densities a_0 and c_0 which satisfy (80) are the *equilibrium* densities.

If on the contrary (as is generally the case) the original densities are not equilibrium densities, *the state is not stationary*; but it *converges* towards a state in which the abundances of the particles (a) and (c) are those given in terms of a_0 and c_0 by Equations (78). However, the rate of convergence (given by $\mathrm{d}c/\mathrm{d}t$) *decreases* rapidly with time, and this explains why an infinite time is needed to reach the stationary equilibrium state.

Many authors immediately deduce from Equation (51) that the 'equilibrium abundances' (defined by $\mathrm{d}c/\mathrm{d}t=0$) are given (from (42)) by (80), but it is not obvious *a priori* that starting from an arbitrary initial distribution, there will be convergence towards equilibrium; and it is even less obvious *a priori* that an infinite time is necessary to attain strict equilibrium.

Remark (Physical Discussion)

In the above calculation, it is not easy to follow explicitly the physical mechanism involved in the convergence towards a stationary state, nor to understand the result completely. Let us therefore consider the following specific case, using the real orders of magnitude of $\tau(a)$ and $\tau(c)$.

We assume that

$$\tau(a) = 80 \times 10^{14}\ \text{sec} = \text{(approximately) } 270 \times 10^6 \text{ years}\,,$$

and

$$\tau(c) = 4 \times 10^{14}\ \text{sec} = \text{(approximately) } 13 \times 10^6 \text{ years}\,.$$

Let $s_0 = 168 \times 10^{14}$ particles per cm^3 be the 'initial' value of the total density. Then from Equations (73), (75), and (74), we have

$$\tau_{\text{cycle}} = \tau(a) + \tau(c) = 84 \times 10^{14}\ \text{sec}\,, \tag{83}$$

$$\begin{aligned} a_\infty = s_0 \frac{\tau(a)}{\tau_{\text{cycle}}} &= 168 \times 10^{14} \times \tfrac{80}{84} = 2 \times 80 \times 10^{14} = \\ &= 160 \times 10^{14}\ \text{particles/cm}^3\,, \end{aligned} \tag{84}$$

$$\begin{aligned} c_\infty = \frac{s_0 \tau(c)}{\tau_{\text{cycle}}} &= 168 \times 10^{14} \times \tfrac{4}{84} = 2 \times 4 \times 10^{14} = \\ &= 8 \times 10^{14}\ \text{particles/cm}^3. \end{aligned} \tag{85}$$

For the 'initial distribution' – that is, the distribution between a_0 and c_0 of the 168×10^{14} particles present in each cm^3 – three typical cases can be considered.

First case. The initial abundances a_0 and c_0 are already the 'equilibrium abundances':

$$a_0 = a_\infty = 160 \times 10^{14} \text{ particles/cm}^3,$$
$$c_0 = c_\infty = \quad 8 \times 10^{14} \text{ particles/cm}^3.$$

We already know that in this case the stationary state will be established immediately at $t=0$ and will be maintained indefinitely. But let us examine more closely the mechanism governing this 'equilibrium'. Let us represent the density schematically by the *width* of a set of parentheses (without any pretensions as to the accuracy of the scale), and let us represent the density itself by the number between the corresponding parentheses. We reserve the parentheses on the left-hand side for the particles (a) and the parentheses on the right-hand side for the particles (c). We indicate the reactions in $dt=1$ second by horizontal arrows 'sub-titled' with the number of particles transformed in one direction or the other.

Now according to the fundamental Equation (36), between $t=0$ and $t=dt=1$ sec we have

$$dn_a/n_a = -\,dt/\tau(a) = -\,1/\tau(a);$$

thus

$$-\,dn_a = a/\tau(a) = a_0/\tau(a) = 160/80 = 2 \text{ particles}. \tag{86}$$

There will be a transfer of 2 particles in the direction

$$(a) \rightarrow (c).$$

Similarly, we have

$$-\,dn_c = c_0/\tau(c) = 8/4 = 2 \text{ particles}. \tag{87}$$

Thus there will also be a transfer of 2 particles in the direction

$$(c) \rightarrow (a).$$

The situation is illustrated by the following diagram:

$$\left(a_0 = 160 \times 10^{14} \text{ particles/cm}^3\right) \underset{2}{\overset{2}{\rightleftarrows}} \left(c_0 = 8 \times 10^{14}\right).$$

Equations (86) and (87) show how, in the case of (dn_a), the overabundance of a compensates for the small transition probability (long lifetime of a); the stationary state is maintained from $t=0$ by the equal number of exchanges in each direction.

In the case under consideration, the distribution at $(t+dt)$ is the same as the initial distribution, and the diagram indicated describes the situation at any time. There is really 'equilibrium' – or, more exactly, a *stationary state.*

Second case. The initial densities a_0 and c_0 are *very different* from the equilibrium abundances (if a_0 is less than a_∞, c_0 is greater than c_∞). Let us take for example:

$$a_0 = \tfrac{1}{2}a_\infty = 80 \times 10^{14} \text{ particles/cm}^3 \qquad c_0 = 11c_\infty = 88 \times 10^{14} \text{ particles/cm}^3.$$

Then, since

$$\tau(a) = 80 \times 10^{14} \text{ sec} \quad \text{and} \quad \tau(c) = 4 \times 10^{14} \text{ sec},$$

we have for $\mathrm{d}t=1$ second and $t=0$:

$$-\mathrm{d}n_a = 1 \text{ particle} \quad \text{and} \quad -\mathrm{d}n_c = 22 \text{ particles}.$$

This time, the diagram takes the form:

$$(a_0 = 80 \times 10^{14} \text{ particles/cm}^3) \underset{22}{\overset{1}{\rightleftarrows}} (c_0 = 88 \times 10^{14} \text{ particles/cm}^3).$$

We clearly see the mechanism for convergence towards the equilibrium state: the underabundance* of (a) *decreases* the loss of these particles (1 instead of 2), while the enormous overabundance* of (c) *increases* the loss of these particles (22 instead of 2). The two phenomena both tend to produce a convergence towards the equilibrium situation. This convergence is rapid at the beginning, in the example under consideration, because of the very large initial 'disequilibrium' $(c_0 = 11c_\infty)$. It would clearly be slower if we were already close to equilibrium, because the net exchange would then be almost zero $(2-2=0)$. This is what occurs in the third case.

Third case. The initial densities are different from the equilibrium abundances, *but the difference is not very marked.* Moreover, for a change, we assume that a_0 is (slightly) greater than a_∞.

Let us take, for example:

$$a_0 = \tfrac{41}{40} a_\infty = 164 \times 10^{14} \qquad c_0 = \tfrac{1}{2} c_\infty = 4 \times 10^{14}.$$

We still have

$$\tau(a) = 80 \times 10^{14} \text{ sec} \qquad \tau(c) = 4 \times 10^{14} \text{ sec}.$$

Then for $t=0$, we will have

$$-\mathrm{d}n_a = 2.05 \qquad -\mathrm{d}n_c = 1.00,$$

and the tendency will again be towards convergence to equilibrium, but with an 'equalization rate' given by $(2.05-1)=1.05$ particles (*slow* convergence).

6. The 'Mean Duration of a Cycle'. The Calculation of the Energy ε when Cyclic Reactions Are Present

We have already learned how to calculate the energy Q' released by *one* cycle of reactions, such as the C-N cycle. Moreover, it is obvious that if R_{cycles} is the number of cycles per cm^3 per second and ϱ is the density of the mixture, the energy released per cm^3 per second $\varepsilon\varrho$ will be given by

* We are concerned with the abundances *with respect to equilibrium proportions*, and not with the abundances themselves.

$$\varepsilon\varrho = Q' \cdot R_{\text{cycles}} . \tag{87}$$

Now, if we assume that the stationary state already exists and if we denote, in a general manner, by (R), (R′), (R″), etc., each of the reactions which make up the 'links' of a cyclic chain, a little thought is enough to convince us that the number of *cycles* per cm^3 per second is equal to the number of times that any *one* of the reactions (R) *or* (R′) *or* (R″), etc. occurs per cm^3 per second. Thus, for example, in the specific case considered in Section 5 (first case), we had

2 reactions (R) of the type $(c) \to (a)$
2 reactions (R′) of the type $(a) \to (c)$

per cm^3 per second. This corresponded to

2 cycles $(c) \to (a) \to (c)$ per second.

This simple example shows that in the stationary state we have

$$\mathrm{R}_{\text{cycles}} = (-\,\mathrm{d}n_a)_{\mathrm{d}t=1} = (-\,\mathrm{d}n_c)_{\mathrm{d}t=1} . \tag{88}$$

In other words, $\mathrm{R}_{\text{cycles}}$ is equal to the number of reactions (R) *or* (R′) per cm^3 per second, whence (using Equation (86) and taking $a_0 = a_\infty$):

$$\mathrm{R}_{\text{cycles}} = a_\infty / \tau(a) = c_\infty / \tau(c) . \tag{89}$$

It is sometimes advantageous to replace the common value of the ratios (89), each of which refers to one of the reactions of the cycle, by a ratio that introduces the total number s_0 of particles per cm^3 and thus the sum of all of the $\tau(a)$, $\tau(c)$, etc. – that is, the quantity that generalizes the definition of τ_{cycle} introduced in Equation (73) to more than two constituents. Then we find

$$\mathrm{R}_{\text{cycles}} = a_\infty / \tau(a) = c_\infty / \tau(c) = \cdots = s_0 / (\tau(a) + \tau(c) + \cdots) = s_0 / \tau_{\text{cycle}} . \tag{90}$$

If $\mathrm{dR}_{\text{cycles}}$ represents the number of cycles taking place during the time $(\mathrm{d}t)$, we will obviously have:

$$\mathrm{dR}_{\text{cycles}} = (\mathrm{R}_{\text{cycles}})\, \mathrm{d}t .$$

And Equation (90) can be written

$$\mathrm{dR}_{\text{cycles}} / s_0 = \mathrm{d}t / \tau_{\text{cycle}} . \tag{91}$$

By analogy with the concept of $\tau_p(c)$ as *'the mean duration of a reaction (R)'* introduced in Section 4.3, we can interpret τ_{cycle} as the *'mean duration of a cycle'*.

We note that when τ_{cycle} is interpreted as the 'mean duration of a cycle', Equation (91) is analogous to (36), which defines the 'mean lifetime' $\tau_p(c)$. If the quantity $(-\mathrm{d}n_c)$ in (36) is interpreted as the *'the number of reactions (R) during the interval* $(\mathrm{d}t)$' and is denoted by $\mathrm{d}n_{\mathrm{R}}$, Equation (36) can be written:

$$\frac{+\,\mathrm{d}n_{\mathrm{R}}}{\text{number of particles } (c) \text{ per cm}^3 \text{ at the time } t} = \frac{\mathrm{d}t}{\tau_p(c)} =$$

$$= \frac{\mathrm{d}t}{\text{'mean duration of a reaction (R)'}}. \tag{92}$$

But (91) can be written in a similar manner:

$$\frac{+\,\mathrm{d}\mathrm{R}_{\text{cycles}}}{\text{number of particles of all kinds per cm}^3 \text{ at the time } t} =$$

$$= \frac{\mathrm{d}t}{\text{'mean duration of a cycle'}}. \tag{91'}$$

When $\tau(a)$ is much greater than $\tau(c)$ (by a factor of at least 100), we have

$$\tau_{\text{cycle}} = \tau(a) \quad \text{(approximately)}. \tag{93}$$

This is what we mean when we say that the rate of energy release is determined by the *slowest* reaction in the cycle. This is the case, of course, not because it is the slowest reaction that contributes the most to Q, but because the order of magnitude of τ_{cycle} is determined by the largest term in the sum $\tau_{\text{cycle}} = \tau(a) + \tau(c) + \cdots$. Actually, the 'slowest' reaction 'determines' the value of $\varepsilon\varrho$ indirectly, for it affects only $\mathrm{R}_{\text{cycles}}$, the number of cycles per cm^3 per second (for a given total density s_0).

7. The Empirical Representation of ε_{pp} and ε_{CN}

The question of the empirical representation, by a simple analytic function, of the curve describing the variation of ε_{12} as given by a relatively complicated function of the type (28) (see the end of Section 3), has lost much of its importance since the development of electronic computers.

Nevertheless, in order to understand certain 'old' but fundamental articles – and, in particular, in order to recover the numerical values of Schwarzschild [1] in the test which we shall apply to his 'model' – we shall find it useful to indicate, if only briefly, the essentials of this method of representation.

Let us denote by ε_{pp} and ε_{CN} the values of ε_{12} for the p-p chain and the C-N cycle, respectively.

The comments made after Equation (87) and the whole of Section 6 show that Equation (28) gives ε_{pp} if we put $X_1 = X_2 = X$ and $Q = \frac{1}{2}Q_{\text{pp}}$. Thus (except for certain well-known numerical constants and other less well-known constants such as S_0 (whose value 'changes' rapidly)), $\varepsilon_{\text{pp}}/\varrho X^2$ depends only on the temperature T of the mixture.

Now, in spite of the complicated analytic form of the relation between ε_{pp} and $T_6 = T/10^6$ (a term in $T_6^{-2/3} \exp[-\beta T_6^{-1/3}]$), it happens that the *logarithm* y of the quotient $\varepsilon_{\text{pp}}/\varrho X^2$:

$$y = \log_{10}(\varepsilon_{\text{pp}}/\varrho X^2) = f(T_6) = F(\log_{10} T_6) = F(x), \tag{94}$$

(lower hatched curve in Figure 13, marked $\varepsilon_{pp}/\varrho X^2$), is a function $F(x)$ that varies 'slowly' enough with $x = \log_{10} T_6$ so that one can approximate relatively long sections of the curve (for long enough intervals in x) as *straight lines*. And for each section, one can put

$$y = Ax + B, \tag{95}$$

where A and B, of course, have different values for each temperature interval.

Following the usual notation, we call $\log_{10}(\varepsilon_{pp})_1$ the constant B (the '*intercept*') and ν the constant A (the *slope of the line* $y = Ax + B$). Thus we arrive at the approximate *analytic* representation (which is 'empirical' in comparison with the curve representing the real function):

$$\log_{10}(\varepsilon_{pp}/\varrho X^2) = \log_{10}(\varepsilon_{pp})_1 + \nu \cdot \log_{10}(T_6). \tag{96}$$

That is, taking the antilogarithm,

$$\varepsilon_{pp} = (\varepsilon_{pp})_1 \varrho X^2 T_6^{\nu}. \tag{97}$$

Schwarzschild [1] gives the following values (Table XIII)* of the constants $\log(\varepsilon_{pp})_1$

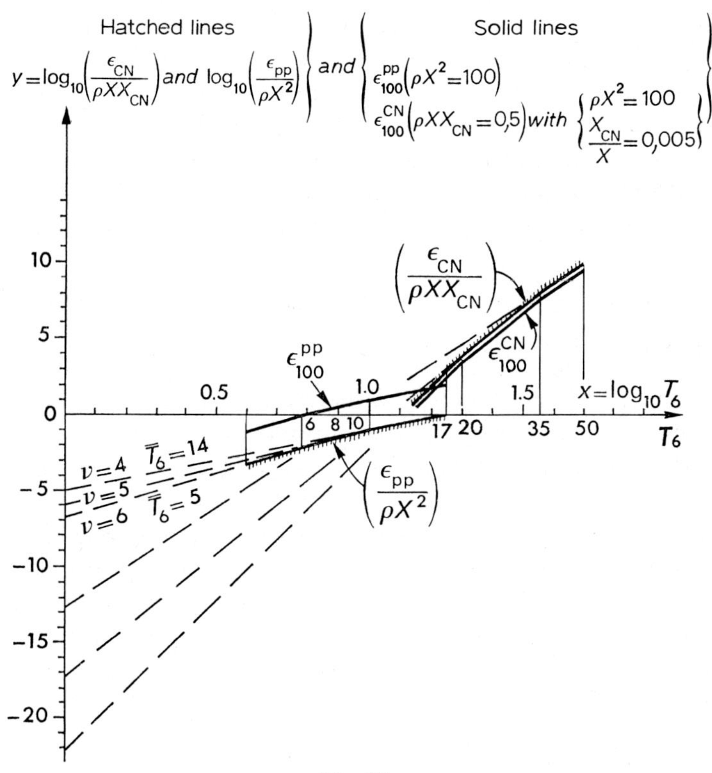

Fig. 13.

* Note that ν *decreases* when the temperature *increases*.

and ν for the range of temperatures found near the center of the Sun (and slightly above it):

TABLE XIII

T_6	$\log_{10}(\varepsilon_{pp})_1$	ν
4–6	−6.84	6
6–10	−6.04	5
9–13	−5.56	4.5
11–17	−5.02	4
16–24	−4.40	3.5

In Figure 13 we have drawn only the sections corresponding to the 'integral' slopes 4, 5, and 6 (broken lines); on a small-scale diagram the difference between half-integral and integral slopes is not perceptible.

In the central regions of the Sun, the usual numerical values of ϱ are of the order of 150 g/cm^3, while X is of the order of 0.50. Thus the product ϱX^2 is of the order of 75 – or, roughly, of the order of 100. Therefore we have also presented in Figure 13 (solid line) the value of ε_{pp} for $\varrho X^2 = 100$. This makes it possible to estimate 'at a glance' the order of magnitude of ε_{pp} for various temperatures. Note that this is not a (poor) '*approximation*' but, fundamentally, a certain *choice of units* which makes possible the rapid evaluation of orders of magnitude.

We also note that the use of integral or half-integral slopes is not a 'miracle'. A curve can always be represented by a tangent of the appropriate 'integral' slope (in the domain where the real slope of the tangents takes on corresponding values); one has only to read from the curve the corresponding intercept, which in general will not be 'integral'.

When we take $\varrho X^2 = 100$, we have (letting ε_{100}^{pp} be the corresponding ε_{pp}):

$$\log_{10} \varepsilon_{100}^{pp} = \log_{10}(\varepsilon_{pp}/\varrho X^2) + \log_{10} 100 = y + 2 . \tag{98}$$

Thus the representation of $\log_{10} \varepsilon_{100}^{pp}$ is derived from that of $y = F(x)$ by 'climbing' *two units* above the line representing the corresponding section of $y = F(x)$. We see in Figure 13 that for T_6 between 6 and 17 $\log_{10} \varepsilon_{100}^{pp}$ varies from 0 to 1.8, which means that ε_{100}^{pp} remains between 1 and 60 ergs g^{-1} sec^{-1}.

In a manner analogous to that just described for the p-p cycle, ε_{CN} can also be represented as a certain number of functions of the type T_6^{ν}. From (28) and (90), ε_{CN} is proportional to $\varrho X X_{CN} = \varrho X^2 (X_{CN}/X)$, where X_{CN} represents the sum of the abundances of C^{12} and N^{14}. Schwarzschild assumes that $X_{CN}/Z = \frac{1}{5}$ (where Z is the abundance of 'heavy elements'). This makes it possible to calculate the function y, which is now defined as

$$\log_{10}(\varepsilon_{CN}/\varrho X X_{CN}),$$

and to represent it as a function of $x = \log_{10} T_6$. This gives the hatched curve located above the axis of $\log_{10} T_6$, and a practical representation of the form:

$$\varepsilon_{CN} = (\varepsilon_{CN})_1 \varrho X X_{CN} T_6^{\nu} . \tag{99}$$

Still following Schwarzschild [1], we have the following values (Table XIV) of $(\varepsilon_{CN})_1$ for various values of ν:

TABLE XIV

T_6	$\log_{10}(\varepsilon_{CN})_1$	ν
12–16	−22.2	20
16–24	−19.8	18
21–31	−17.1	16
24–36	−15.6	15
36–50	−12.5	13

In order to find the order of magnitude of ε_{CN} under the conditions prevailing in the central regions of the Sun, we shall take, as before, $\varrho X^2 = 100$ with $Z = 0.020$. With $X_{CN} = Z/5$ and $X = 0.50$, this gives us $X_{CN}/X =$ (approximately) 0.008. Whence

$$\log_{10}(\varrho X X_{CN}) = \text{approximately } (-0.1) .$$

Thus the curve representing $\log_{10} \varepsilon_{100}^{CN}$ is obtained by lowering the corresponding hatched curve by 0.1 units. Thus we see that when the CN cycle operates, it releases energies whose logarithm varies from 2 to 9. Therefore the energy release varies from 100 to 10^9 ergs g^{-1} sec^{-1}.

But for the Sun, where T_6 is less than 15, it is the p-p cycle that dominates.

For calculations using only first-order approximations, one often takes $\nu = 4$ for the whole of the p-p cycle in the Sun (a mean T_6 of 14), and $\nu = 18$ for the whole of the C-N cycle (a mean T_6 of 20).

As we see in Figure 13, the choice of $(\varepsilon_{pp})_1$ and of $(\varepsilon_{CN})_1$ makes it possible to compensate quite easily for a poor choice of ν. The essential factor is to take the $(\varepsilon_{pp})_1$, or the $(\varepsilon_{CN})_1$, which corresponds to the value adopted for ν.

Remark. We should not be surprised by the comparison between the values of ε_{100}^{pp} and ε_{100}^{CN} from Figure 13, and the well-known value:

$$L_{\odot}/M_{\odot} = 2 \text{ ergs g}^{-1} \text{ sec}^{-1} .$$

This latter value refers to the total mass of the Sun, while ε_{100}^{pp} and ε_{100}^{CN} are relevant only in the 'core' (r' less than 0.35) of the Sun, where the temperature is high enough for thermonuclear reactions to take place.

8. Application to the Sun. The 'Final Test'

8.1. Review of the Principal Results

The study of the relation between the net output of radiation L_r and the temperature gradient dT/dr, carried out in Chapter VII of [9], and the study of the energy equilibrium (nuclear reactions) which we have just completed, have resulted in a long digression, after which we should review the situation by recalling where we were at the end of our Chapter III.

Starting from a certain density distribution $\varrho(r')$, which is in principle arbitrary but which is actually chosen in such a way as to satisfy the set of conditions defining a certain model of the Sun at a certain stage in its evolution (the Schwarzschild model with a *central* hydrogen abundance equal to 0.50), we have shown (Table II) how the integration of the equation $\mathrm{d}M'/\mathrm{d}r' = 2.13 \times \varrho r'^2$ enabled us to obtain the function $M'(r')$. Next, the use of the equation of mechanical equilibrium

$$\mathrm{d}P/\mathrm{d}r' = -\,1.91 \times 10^{15}\, M'\varrho/r'^2$$

(integrated from the surface towards the interior) gave us the values of the function $P(r')$ – and, in particular, the value $P(0)$ of the total pressure at the center.

We then used the approximate relation

$$P = P_{\mathrm{gas}} + P_{\mathrm{rad}} \approx \frac{k}{m_{\mathrm{H}}}\,(0.75 + 1.25\,X_0)\,\varrho T\,,$$

which was obtained by neglecting P_{rad} in comparison with P_{gas}, by neglecting the variation of $X(r')$ in the neighbourhood of the center, and by taking $X(r') = X(1) = X_0 = 0.744$. This gave us a table of values of the function $T(r')$ which is accurate in the region where these approximations are valid – that is, the region between the surface and $r' = 0.3$.

It now remains for us to show, from Equations (97) and (99) of the preceding section, that the value of $\varepsilon_r = \varepsilon_{\mathrm{pp}} + \varepsilon_{\mathrm{CN}}$ is actually *zero* in the region in question (r' greater than 0.3), for the values of $\varrho(r')$, $X = X_0$, and $T(r')$ assumed according to the preceding argument (this will be the purpose of Table XVII below). The fact that $\varepsilon_r = 0$ implies that L_r (or L_r') is *constant* in the corresponding region. In this region we then have $L_r' = 1$. But we must still satisfy the relation obtained in [9], by a study of the opacity; this relation is repeated in Equation (100) below. As unknown functions it contains only $T(r')$ and $\varrho(r')$. Our choice of $\varrho(r')$ will be definitively justified if we find that this relation is satisfied for r between 0.86 and 0.30; for we have already seen in Section 6 of Chapter III that in the convective zone the 'convective' relation between P and T is satisfied.

8.2. The Region in Which ε Is Negligible*

First we must show that the values of $\varrho(r')$ in Table I do lead to a function L_r which is independent of r in the region ($r' > 0.3$) where the production of nuclear energy is negligible and where ($r' < 0.86$) energy is not yet transported by convection.

In principle, we have only to verify that Equation (116) of Chapter VII in [9], where we have put

* Before continuing the study of this section, the reader should become familiar with (or review) Chapter VII of *Introduction to the General Theory of Particle Transfer* [9] by the present author, which deals with the mechanical effects of radiation (radiation pressure) and with the concept of opacity.

$$X = 0.744 = X\,(r' = 1) = \text{const}$$

or

$$L'_r = -\,4.46 \times 10^{-49}\, r'^2 T^{7.5} \varrho^{-1.75}\, \mathrm{d}\,(\log_{10} T)/\mathrm{d}r' \,, \tag{100}$$

does give $L'_r = 1.00 = \text{const}$ with the function $T(r')$ which corresponds, through $P(r')$, to the distribution adopted in Table I for $\varrho(r')$.

However, with the interval $\Delta r' = 0.10$ adopted in our calculations, the values of $\mathrm{d}(\log_{10} T)/\mathrm{d}r'$ deduced from the values of $T(r')$ would not be precise enough, and it is better to perform this test by inverting Equation (100) – that is, by calculating $\mathrm{d}(\log_{10} T)/\mathrm{d}r'$ with this formula, taking $L'_r = 1.00$. We will then have to verify that the derivatives thus obtained give tangents to the curve representing $\log_{10} T$ which satisfactorily 'envelop' this curve, constructed with the preceding values of $\log_{10} T$. We thus obtain the calculation indicated in Table XV.

TABLE XV

r'	$\log_{10} T_6$	$T^{7.5}{}_{\text{exact}}$	$[\mathrm{d}\log_{10} T_6/\mathrm{d}r']_{\text{calc}}$	$[\mathrm{d}\log_{10} T_6/\mathrm{d}r']_{\text{exact}}$
0.80	0.103	5.92×10^{45}	-1.72	-1.78
0.70	0.256	8.39×10^{46}	-1.41	-1.42
0.60	0.397	9.73×10^{47}	-1.32	-1.37
0.50	0.535	1.02×10^{49}	-1.39	-1.40
0.40	0.676	1.19×10^{50}	-1.41	-1.42
0.30	0.823	1.49×10^{51}	-1.47	-1.48

The last column of Table XV gives the more precise values of $\mathrm{d}(\log_{10} T_6)/\mathrm{d}r'$ obtained by taking $\Delta r' = 0.02$ and by carrying more decimals in the calculations.

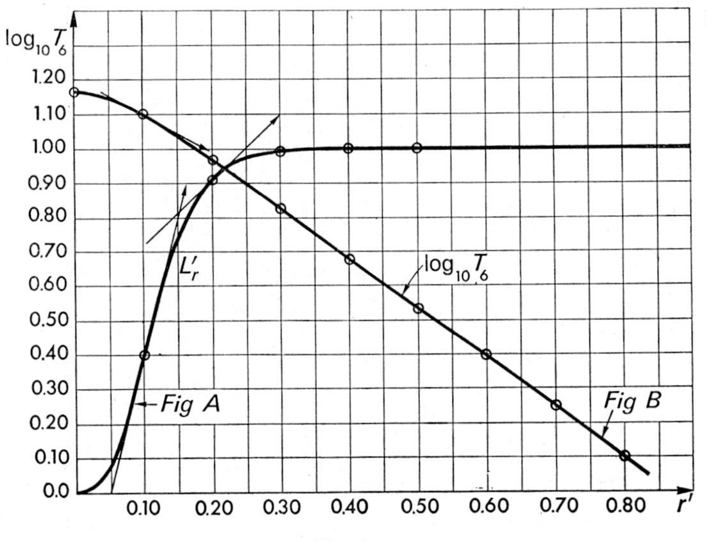

Fig. 14.

Representing the function $\log_{10} T_6$ as a function of r' on graph paper, with the scale $\Delta r' = 0.10 \sim 20$ mm and $\Delta \log_{10} T_6 = 0.10 \sim 10$ mm, we find that the test succeeds perfectly (Figure 14B).

8.3. The 'Central' Region ($r' < 0.40$) Where L_r' and X Vary

For r' less than or equal to 0.40, we must have not only $\varrho(r')$, which by itself determines the values of M_r and of $P(r)$, but also the function $X(r)$, without which we cannot have either the precise value of μ (which is necessary for the calculation of T from P) or the value of ε_r (which is proportional to X^2). But the choice of these two functions should be justified by passing the following 'test'.

Let us calculate $T(r')$ from $\varrho(r')$ and the corresponding values of $P(r')$ by the formula $T = (\mu m_{\mathrm{H}}/k)P/\varrho$, determining μ from the values adopted for $X(r')$ and the formula $1/\mu = 0.75 + 1.25X$. We thus obtain the values of T_6 given in the last column of Table XVI, where the third column gives the values adopted for the function $X(r)$.

With the values of $T(r')$ obtained in this way and repeated in the first column of Table XVII, let us calculate the function ε_r (with the same adopted values for $X(r')$) using the formulae (established in Section 7):

$$\varepsilon_r = \varepsilon_{\mathrm{pp}} + \varepsilon_{\mathrm{CN}} = \varepsilon_{\mathrm{pp}}^1 \varrho X^2 T_6^\nu + \varepsilon_{\mathrm{CN}}^1 \varrho X X_{\mathrm{CN}} T_6^\nu .$$

TABLE XVI

r'	ϱ	X	μ	P_{exact}	T_6^{exact}
0.0	135	0.494	0.734	2.24×10^{17}	14.62
0.1	86	0.611	0.660	1.36×10^{17}	12.65
0.2	36.4	0.723	0.606	4.62×10^{16}	9.35
0.3	13.0	0.744	0.595	1.18×10^{16}	6.65
0.4	4.1	0.744	0.595	2.71×10^{15}	4.74

TABLE XVII

r'	T_6	$\varepsilon_{\mathrm{pp}}^1 T_6^\nu$ / $\varepsilon_{\mathrm{CN}}^1 T_6^\nu$	ϱX^2 / $\varrho X X_{\mathrm{CN}}$	$\varepsilon_{\mathrm{pp}}$ / $\varepsilon_{\mathrm{CN}}$	ε_r
0.0	14.62	0.436 / 12.6	33.0 / 0.17	14.4 / 2.1	16.5
0.1	12.65	0.245 / 0.69	32.1 / 0.2	7.9 / 0.2	8.1
0.2	9.35	0.065 / –	19.0 / –	1.2 / –	1.2
0.3	6.65	0.012 / –	7.2 / –	0.08 / –	0.1
0.4	4.74	– / –	– / –	– / –	0.0

We then calculate dL'/dr' by applying the relation $dL_r/dr = \varepsilon_r\, dM_r$ in the form

$$dL'/dr' = 1.11\ \varrho r'^2 \varepsilon_r . \tag{101}$$

We obtain the column headed dL'/dr' in Table XVIII.

Integrating these values of dL'/dr' numerically from $L'(0)=0$, we should in principle find the values of $L'(r')$ given in the last column of Table XVIII.

TABLE XVIII

r'	ϱ	ε_r	dL'/dr'	L'_{ex}
0.0	135	16.5	0.00	0.000
0.1	86	8.1	7.75	0.396
0.2	36.4	1.2	1.9	0.909
0.3	13.0	0.1	0.1	0.994
0.4	4.1	0.0	0.0	1.000

Remarks. Actually, the last column of Table XVIII does not give L' as obtained by the integration of the values of dL'/dr' in the preceding column, for the function $L'(r')$ varies much too quickly in this region for $\Delta r'=0.1$ to be a satisfactory interval in the calculations. What we give is the result of a calculation made with the interval $\Delta r'=0.02$. Nevertheless, we see in Figure 14A that our values of dL'/dr' give tangents in agreement with the curve representating $L'_{\text{exact}}(r')$.

For $r'=0$, we have

$$dL'/dr' = 4\pi\varrho_c r_c^2 \varepsilon_{r,\,c} R_\odot/L_\odot \to 0,$$

because $r_c=0$. Thus the tangent at the origin is horizontal.

Note the rapidity with which L' increases in the neighbourhood of $r'=0.1$, where $dL'/dr'=7.75$.

8.4. The 'Final Test'

The function $L'(r')$ is connected to the temperature $T(r')$ not only by the calculations indicated in Tables XVI, XVII and XVIII, but also by a relation involving the opacity – that is, Equation (116) in Chapter VII of [9] – which can be written in another form (avoiding the specification of X as in (100)) by solving for $d(\log_{10} T_6)/dr'$. With $C_0 = 1.48\times 10^3$, this gives

$$-\,d\log_{10} T_6/dr' = C_0 r'^{-2} T_6^{-7.5} \varrho^{1.75} (1+X)^{0.75} L'_r . \tag{102}$$

If this equation gives the same function $L'(r')$ as that obtained in Table XVIII, we can consider that $\varrho(r')$ and $X(r')$ have both been well chosen. However, for the same reasons as in Section 8.2, it is convenient to calculate the derivative of $\log_{10} T_6$ by Equation (102) *using* the function $L'(r')$ given in Table XVIII. We will have passed the test if we recover in this way the ('exact') value of $d(\log_{10}T_6)/dr'$ given by the relation $T = (\mu m_H/k)P/\varrho$, and thus obtained from $\varrho(r')$ alone.

Table XIX gives the details of the corresponding calculation.

TABLE XIX

Calculation of $d(\log_{10} T_6)/dr'$ by means of Equation (102)

r'	$T_6^{-7.5}$	$(1+X)^{0.75}$	L'_{ex}	$d(\log_{10} T_6)/dr'$
0.0	1.82×10^{-9}	1.354	0.000	0.00 [a]
0.1	5.61×10^{-9}	1.43	0.396	−1.13
0.2	5.20×10^{-9}	1.50	0.909	−1.41
0.3	6.78×10^{-7}	1.517	0.994	−1.48
0.4	8.50×10^{-6}	1.517	1.000	−1.42

[a] For $r' = 0$, ϱ and ε_r can be considered constant in the equation $dL_r/dr = 4\pi\varrho\varepsilon_r r^2$. Thus in the neighbourhood of $r' = 0$, L_r' varies as $(r')^3$.

As usual we can dispense with the laborious calculation of the 'exact' logarithmic gradient by simply verifying that the approximate $d(\log_{10} T_6)/dr'$ given by (102) yields correct tangents to the curve of $\log_{10} T_6$ as a function of r'. A single diagram (Figure 14B) is sufficient to carry out the test proposed here and the test which was to be performed at the end of Section 8.2.

CONCLUSION. *The values of* $\varrho(r)$ *and* $X(r)$ *given by Schwarzschild* [1] *satisfy all the conditions of the problem.*

Exercise 8

(1) In an astrophysical text dating from 1928, James Jeans writes: "The equation which expresses the fact that the radiative flux is determined by the quantity of energy produced each second by unit mass in the interior of the star, can be written:

$$-4\pi r^2 (4aT^3 C/3k\varrho)(dT/dr) = \int_0^r 4\pi r^2 G \, dr." \quad (1)$$

Knowing that Jeans' notation is not exactly the same as modern notation, you are asked:

(1a) To prove (*given* the fundamental differential equations of the theory of stellar interiors) Equation (1′), which is physically identical (except for notation) with (1).

(1b) To give the equivalents of Jeans' quantities C, G, and k, in the usual notation. What 'name' should be given to Jeans' C, G, and k in order to indicate precisely their physical nature?

(2) Jeans writes, in the same work: "The heat flux per cm² per second, which corresponds to a temperature gradient $\partial T/\partial r$, in the case where heat is transferred by radiation, is given by the equation

$$\pi F = -\tfrac{1}{3}\varrho C_v (\partial T/\partial r)\, \bar{c}\lambda, \quad (2)$$

where C_v is the specific heat at constant volume (in ergs per gram per degree), $\bar{c}$ is the mean velocity and λ the mean free path of the particles (photons)."

(2a) Compare Equation (2) with the classical expression for πF, so as to find expressions for Jeans' C_v, $\bar{c}$ and λ in terms of the usual quantities with which they are most directly connected. [You can help yourself, if necessary, by considering (1).]

(2b) Jeans adds that Equations (1) and (2) are only approximate. Justify this statement by briefly enumerating all the approximations implied by each equation.

(3) Jeans then gives the equation

$$\mathrm{d}P/\mathrm{d}r = -(\gamma\varrho/r^2)\int_0^r 4\pi\varrho r^2\,\mathrm{d}r, \tag{3}$$

and says: "The physical meaning of this equation is of course that the total pressure P is just adequate to support the weight of the whole column of gas standing above it." Then, setting

$$\bar{G}\cdot M_r = \int_0^r G\cdot \mathrm{d}M_r \tag{5}$$

and

$$M_r = \int_0^r 4\pi\varrho r^2\,\mathrm{d}r, \tag{5}$$

he states that in a star in mechanical and energy equilibrium, we have

$$(4\pi C\gamma/k\bar{G})\,(\mathrm{d}P_{\mathrm{rad}}/\mathrm{d}r) = (\mathrm{d}/\mathrm{d}r)\,(P_{\mathrm{gas}} + P_{\mathrm{rad}}). \tag{6}$$

(3a) What does γ stand for in Equation (3) and in Equation (6)?

(3b) Are you satisfied with the physical interpretation Jeans gives to Equation (3)?

(3c) What is the physical meaning of $\bar{G}$? Does this quantity generally depend on r? What more modern notation do you propose for $\bar{G}$?

(3d) Prove the 'modern' equivalent of Equation (6), using the usual differential equations and the usual constitutive relations.

(4) Jeans says: "Equations (4) and (6) were first used in 1917 by Eddington, who assumed that the product $k\bar{G}$ is independent of r."

(4a) Indicate the three fundamental parameters of internal structure through which Jeans' k (see the result of 1b) depends, in general, on r.

(4b) Show that the 'Eddington hypothesis' leads to a relation between ϱ and T of the form

$$\varrho = \varrho_0 T^n, \tag{7}$$

where ϱ_0 is a quantity independent of r. What is the numerical value of n? What is the polytropic index of the corresponding model?

(4c) According to Jeans, ϱ_0 is given by the equation:

$$\varrho_0 = (am\mu/3R)\,[(4\pi C\gamma/k\bar{G}) - 1]. \tag{8}$$

Comparing the expression for ϱ_0 in the usual notation with (8), find the meaning of R and m; a, $\bar{G}$, k, and γ have the same meaning as before, while μ has its usual meaning.

(5) Finally, Jeans says that the absorption coefficient k is of the form

$$k = c'F\varrho T^{-3.5}/\mu(1+f), \tag{9}$$

where F is "the ratio of the total absorption to that caused by free electrons alone", and $(1/f)$ is "the ratio between the number of free electrons and the number of free nuclei which would be present if there were complete ionization".

Assuming that μ has its usual meaning, you are asked:

(5a) To show that $\mu(1+f)$ can be very simply expressed in terms of the ratio (ϱ/N_e), where ϱ is the density of the mixture and N_e is the number of free electrons per cm^3. (Write this proof without bringing in the composition X, Y, Z, of the mixture.)

(5b) To find an expression for $1/\mu(1+f)$ in terms of the hydrogen abundance X.

(5c) Taking Jeans' notation into account, is Equation (9) in agreement with the formula given by 'modern' calculations of the Rosseland mean? (Assume that the *result* of these calculations is known, and do not re-compute it. See [9], Chapter VII.)

(6) According to Schwarzschild's model for the present state of the Sun, the quantity k_J which Jeans denotes by k takes on the values given in the following table:

r'	0.0	0.1	0.2	0.3	0.4	0.5	0.6	0.7	0.8
$100k_J$	107	134	212	323	451	589	748	959	1290

In the light of this information, does the 'Eddington hypothesis' seem reasonable to you?

(6a) How do you explain the fact that the 'Eddington hypothesis' gives a value of n (in Equation (7)) that is quite correct in order of magnitude?

Exercise 9

N.B. – Give all the details of the calculations, which are to be performed with a 4-place logarithm table or with a slide rule.

We propose to study the *exterior* regions (E) of a star (S), composed of homogeneous concentric layers. The radius of the star is R, its luminosity is L, and its mass is M. We assume that the total mass of the 'envelope' (E) is negligible in comparison with M, and that there are no nuclear reactions in (E). We shall also neglect the radiation pressure P_{rad} in comparison with the gas pressure, except in the last question (R 11).

In order to shorten the expression of certain functional dependences, we shall use the 'universal' symbol $f(...)$ to denote any relation among certain variables, all of which need not necessarily be mentioned explicitly; $f(...)$ need not always represent the same function.

Let (I), (II), (III), (IV) be the classical expressions for dM_r/dr, dP/dr, dL_r/dr, and the relation between L_r and dT/dr. Let (V) be the perfect gas law, with the gas constant written in the form (k/H) to avoid confusion with R. Let (VI) be the expression for P_{rad} in terms of T. Let (VII) be the expression for the opacity $\bar{k}$ in cm² per cm³. We recall that $a = 7.6 \times 10^{-15}$, $G = 6.67 \times 10^{-8}$, and $k/H = 8.32 \times 10^7$ c.g.s. In $\bar{k}$, the exponent of ϱ will be taken to be an *integer*!

(1) Replace Equations (II) (III) and (IV) with a set of simpler relations resulting from the particular conditions assumed above. (Think carefully about this 'substitution', for a mistake here will affect all the rest of the problem!) Explain as clearly as possible why the same modification should not be applied to Equation (I). Explain why the fact of neglecting the radiation pressure (as we have said above) does not allow us to neglect Equation (IV), even when we express it, as Eddington did, in the form which introduces the gradient of P_{rad}. What factor appearing in $\bar{k}$ remains constant in (E)?

We shall then distinguish two cases: case (C) for a 'convective' envelope and case (R) for an envelope in radiative equilibrium.

Case (C)

In this case we have

$$dT/dr = (2T/5P)(dP/dr).$$

(C2) Formulate and integrate the differential equation which directly connects the variables T and P, without explicitly containing the variable r. Show that the usual boundary conditions are perfectly compatible with an *arbitrary* integration constant, which we shall call B. Let (1) be the relation $T = f(P, B)$ obtained in this way.

(C3) Next find an equation of the form $dP/dr = f(P, r)$, where f is independent of ϱ. Integrate this equation. The constant of integration of this new equation necessarily depends on B, even after the boundary conditions are applied. Let (2) be the solution $P^{2/5} = f(r, B)$. Let (3) be the corresponding expression $T = f(r)$. What fortunate simplification is found when going from (2) to (3)? Could it have been foreseen?

(C4) Calculate for the Sun

$$(M = 1.98 \times 10^{33} \text{ c.g.s.} \quad L = 3.78 \times 10^{33} \text{ c.g.s.}$$
$$R = 6.95 \times 10^{10} \text{ c.g.s.} \quad X = 0.744 \text{ in } E),$$

the value of the constant of proportionality between T and $z' = (1 - r')/r'$ in (3). Then determine T for $r' = 0.98 - N \times 0.04$, where $N = 0, 1, 2, 3, ...$ up to the value at which the above equations are no longer applicable. Are the values thus found for T compatible with the hypotheses made in this calculation?

(C5) Does transfer theory make it possible to determine the value of B when we assume the existence of an '*atmosphere*' in radiative equilibrium above (E)?

Case (R)

(R6) Redo the calculation of (C2) by first formulating an equation of the form $\mathrm{d}P/\mathrm{d}T = f(P, T)$ and integrating it; after the determination of the integration constants by means of the boundary conditions, this gives an Equation (4) of the form $P = f(T)$. We shall denote by A the constant factor which depends on G, M, a, c, k, H, X, μ, and L on the right-hand side of (4).

(R7) Formulate the equation $\mathrm{d}T/\mathrm{d}r = f(T, r)$ directly, without using the explicit calculation of $P = f(r)$. Is it appropriate to set $\mathrm{d}T/\mathrm{d}r = f(T, r)$, considering the result obtained by trying to put $\mathrm{d}T/\mathrm{d}r$ in this form? Show that the final integration still gives a solution of the form $T = f(z')$. We shall call this solution T_{rad}, to distinguish it from the corresponding solution T_{conv} in case (C).

(R8) Show that there exists a very simple *mathematical* relation between T_{rad} and T_{conv}. Does this relation have a physical meaning? Could we have foreseen, *a priori*, the existence of such a relation?

(R9) Is the fact that $\mathrm{d}T_{\mathrm{rad}}/\mathrm{d}r$ is independent of L due to a fortuitous balance, or does it correspond to some simple physical reason?

(R10) Apply the 'stability' criterion to the relation between $\mathrm{d}(\log P)/\mathrm{d}r$ and $\mathrm{d}(\log T)/\mathrm{d}r$ in case (R), to see whether case (R) can be realized in certain stars. The reply can be only 'always yes' or 'always no'. However, there are stars whose envelopes are 'convective' and stars (hot and massive) whose envelopes are 'radiative'. How do you explain this apparent contradiction? (This question is particularly difficult; do not spend too much time on it.)

(R11) The existence of envelopes in radiative equilibrium in certain hot stars where the radiation pressure is not negligible, leads us to reexamine the question without neglecting P_{rad}. We shall do this by a method of Chandrasekhar, which consists of introducing a system of auxiliary variables:

$$r' = r/R \qquad T' = T/T_0 \qquad \varrho' = \varrho/\varrho_0 \qquad x = K\sqrt{T'} \qquad y = K\varrho'/(T')^3,$$

where

$$\varrho_0 = \alpha^8\beta^7 \qquad T_0 = \alpha^2\beta^2 \quad \text{and} \quad K = (k/H)\, 3\alpha^2\beta/a\mu,$$

with

$$\alpha = 4\pi cGM/k_0L \quad \text{and} \quad \beta = 3GM/aR.$$

(Here k_0 represents that part of the opacity formula which is independent of ϱ and T.)

(R11a) Find the (very simple) physical meaning of the variable y.

(R11b) Following the calculations already performed at the beginning of the problem, find the very simple first-order differential equation (which is not so easy to solve) satisfied by the function $y(x)$. Verify that it is of the form

$$\mathrm{d}(y^2)/\mathrm{d}x = f\,[y(y+1)/x].$$

(R11c) Show that this equation reduces to

$$\mathrm{d}(y^2)/\mathrm{d}x = f\,(y^2/x),$$

if we neglect the radiation pressure.

(R11d) This 'reduced' equation is easy to solve. Integrate it, taking into account the classical boundary conditions. Returning to the variables T and z', show that the result $T_{\mathrm{rad}} = f(z')$ of (R7) is recovered in this way.

(R11e) Suggest a method for integrating the complete equation of (R11b.)

Start the integration, to a first approximation.

Exercise 10

Some calculations can be performed with a slide rule (used to its maximum precision). Others must be carried out with 5-place logarithm tables. All the data are in c.g.s. units. The quantities between brackets are the logarithms (to base 10) of the corresponding numbers. To show that a value Q is expressed as a function of the variables T, P, ϱ, r', etc., and of the parameters μ, X_0, etc., we shall write

$$Q = f(T, P, \varrho, r', \ldots; \mu, X_0).$$

Numerical data. The meaning of certain symbols will be specified later *(in the logarithms use only 4 decimal places)*:

$$G = 6.668 \times 10^{-8} = [\bar{8}.8240]$$
$$k/m_{\mathrm{H}} = R' = 8.317 \times 10^{+7} = [7.9200]$$
$$a/3 = 2.520 \times 10^{-15} = [\overline{15}.4014]$$
$$M = 1.989 \times 10^{+34} = [34.2986]$$
$$R = 4.238 \times 10^{+11} = [11.6272]$$
$$L = 1.973 \times 10^{+37} = [37.2951]$$
$$\bar{\varrho} = 0.0624 = [\bar{2}.7952] \qquad t = 34 \times 10^{9} \text{ years} \qquad X^{\odot}_{\mathrm{CN}} = Z_0/3$$
$$X_0 = 0.90 \qquad Y_0 = 0.09 \qquad Z_0 = 0.01$$
$$r'_1 = 0.160 \qquad r'_2 = 0.0786;$$
$$3L \log_{10} e/64\pi\sigma R = 5.33 \times 10^{27} \qquad GM/R = 3.129 \times 10^{15}$$
$$4\pi R^3/L = 0.0483 \qquad \varepsilon_r = (8/3) \times 10^{-20} X\varrho T_6^{16}.$$

In his book *Structure and Evolution of the Stars*, Martin Schwarzschild (whom we shall call MS) considers a stellar model of mass M, whose radius at age t equals R, and whose corresponding luminosity is L. The model consists of three 'zones' (S), (I), and (C).

The zone (S) extends from $r' = 1.00$ to

$$r' \approx 0.16 = r'_1.$$

It is characterized by a homogeneous composition X_0, Y_0, Z_0, X^0_{CN} and is in radiative equilibrium. The 'intermediate' zone (I) is also in radiative equilibrium, but here X and Y vary with r' while $Z = Z_0$ and $X_{\mathrm{CN}} = X^0_{\mathrm{CN}}$. The central zone ($C$) is in convective equilibrium and its composition is homogeneous, with $X = X_c = 0.061$ and $Z = Z_c = Z_0$. The zone (C) extends up to $r' = r'_2 = 0.0786$. MS gives a table (T) for this model, reproduced below.

(1) Locate M and R approximately with respect to the mass and the radius of the Sun. What

TABLE T

r'	$L' = L_r/L$	$10^3 X$	$\log(P/10^{15})$	$\log(T/10^7)$	$\log \varrho$	$\bar{k}/\varrho$	$M' \times 10^3$
0.00	0.000	61	1.844	0.545	1.075	conv.	
0.02	0.131	61	1.824	0.537	1.063	–	1.50
0.04	0.582	61	1.764	0.513	1.027	–	11.4
0.06	0.974	61	1.662	0.473	0.966	–	35.5
0.08	1.000	81	1.506	0.415	0.868	0.417	75.1
0.10	1.000	325	1.223	0.357	0.648	0.415	119
0.12	1.000	531	0.998	0.310	0.475	0.414	158
0.14	1.000	720	0.826	0.268	0.347	0.416	203
0.16	1.000	893	0.670	0.231	0.231	0.418	245
0.18	1.000	900	0.575	0.196	0.172	0.426	289
0.20	1.000	900	0.478	0.162	0.111	0.434	337
0.22	1.000	900	0.379	0.129	0.046	0.442	388
0.24	1.000	900	0.276	0.096	−0.022	0.451	441
0.26	1.000	900	0.170	0.063	−0.095	0.459	493
0.28	1.000	900	0.061	0.030	−0.171	0.468	544
0.30	1.000	900	$\bar{1}.949$	$\bar{1}.998$	−0.249	0.476	595
0.40	1.000	900	$\bar{1}.351$	$\bar{1}.835$	−0.683	0.516	796
0.42	1.000	900	–	$\bar{1}.802$	−0.776	0.524	826
0.50	1.000	900	$\bar{2}.704$	$\bar{1}.670$	−1.166	0.552	914
0.60	1.000	900	$\bar{2}.002$	$\bar{1}.497$	−1.695	0.587	970
0.70	1.000	900	$\bar{3}.213$	$\bar{1}.306$	−2.292	0.626	992
0.80	1.000	900	$\bar{4}.236$	$\bar{1}.070$	−3.034	0.682	999
0.90	1.000	900	$\bar{6}.762$	$\bar{2}.716$	−4.157	0.777	1000
1.00	1.000	900	–	–	–	–	1000

are the principal differences between the structure of the model in question and the present structure of the Sun?

(2) In which 'zone' should we have $y(r') = \log T - (0.4) \log P$ constant (independent of r'), if $\gamma = 5/3$? Is this 'test' satisfied by the values of Table (T)?

(3) Indicate, *without any calculations*, all the peculiarities of Table (T) which reflect the principal properties of the separation of the star into zones (S), (I) and (C) as defined above. How do you explain the fact that $L' = 1.0000$ in a zone where X varies with r'?

(4) Test the accuracy of Table (T) by calculating $\mathrm{d}(\log_{10} T)/\mathrm{d}r' = f(\bar{k}, L', T, r')$ for $r' = 0.08$, 0.10, 0.40 and 0.42, and by comparing the corresponding values of the variation of $\log_{10} T$ for $\Delta r' = 0.02$ with those obtained from Table (T).

(5) Calculate the values of $\mathrm{d}L'/\mathrm{d}r' = f(\varrho, T, r')$ from the values of ϱ and T in Table (T), for $r' = 0.04$, $r' = 0.06$, $r' = 0.10$ and $r' = 0.12$, and apply the 'test' to the corresponding values of the variation of L' for $\Delta r' = 0.02$ in the appropriate interval. Does the result (assuming that it is correct) partly explain the corresponding values of L' given by (T)?

(6) For all the values of r' in Table (T) located in the zones (C) and (I), calculate the values of the gas pressure P_{gas} and the radiation pressure P_{rad}. Give the answer in the form of a table of ($P_{gas}/10^{16}$) and ($P_{rad}/10^{16}$) to 0.001. *Compare your result with the fourth column of Table* (T), after having deduced from your P_{gas} and P_{rad} the value of $\log(P/10^{15})$. Find the values of the difference

$$\Delta(r') = \log P - \log P_{calc},$$

where P_{calc} is the total pressure obtained from *your* calculation and P represents, as before, the values in Table (T). (Find this difference Δ only for the zones C and I; round off your values of Δ to 0.01). Complete your calculation of $\Delta(r')$ by determining $\Delta(0.18)$ and $\Delta(0.20)$.

(7) What systematic oversight of MS could have produced the very disappointing appearance of Δ between $r' = 0.00$ and $r' = 0.16$?

(*N.B.* – The calculations in (8) are to be performed with a slide rule!)

(8) To check the values of P given by Table (T), we shall assume that the values of ϱ are *all* correct (in which case the values of M' will also be correct, but you are not asked to 'test' this statement; you may accept it without proof). By integrating the equation $\mathrm{d}P/\mathrm{d}r' = f(M', \varrho, r')$ from the surface inwards, determine the values of $P(r')$ for r' going from 1.00 to 0.30 in steps of $\Delta r' = 0.10$; then for r' going from 0.30 to 0.00 in steps of $\Delta r' = 0.02$. (This calculation is not very long if one is careful not to exceed a precision of 0.01 in $(\mathrm{d}P/\mathrm{d}r') \times 10^{-17}$.) Compare the result with the values of P given in Table (T). For what values of r' does the disagreement for the function $P \times 10^{-16}$ exceed 0.01? Thus if the ϱ and M' given by MS are accurate, as we provisionally assume, the non-zero value of $\Delta(r')$ in certain regions can arise from the inaccuracy of the 4th column of (T) alone; $\log T$ need not necessarily be incorrect in the same region.

(9) Let $P_\varrho(r')$ be the total pressure as deduced in (8) from the values of ϱ.

Show that the corresponding central temperature is not very far from 33.66×10^6 degrees, and determine this central temperature $T_{\varrho,c}$ by determining the central pressure which corresponds to 33.6×10^6 and 33.8×10^6 K. Deduce from your value of $T_{\varrho,c}$ the temperatures T_ϱ for $r' = 0.02$, 0.04 and 0.06, taking advantage of the principal physical characteristic of each zone. Is the difference between $\log T$ from Table (T) and $\log T_\varrho$ for these values of r' very different from 0.020? Is it constant or variable with r'? Where could this difference come from?

(10) What conclusions do you draw from your calculations concerning Table (T)?

CHAPTER V

EVOLUTIONARY MODELS. THE ACTUAL DETERMINATION OF STRUCTURE

1. Introduction

In the four preceding chapters we have tried to establish the *physical relations* which the principal 'structural parameters' (such as M_r, P, T, and L_r) must satisfy and we have also attempted to demonstrate their 'interrelationships' and their orders of magnitude for different values of $r' = r/R$ (especially at the center and at the surface of the star).

This enabled us to state precisely the specific role played by each of the four fundamental differential equations [in $\mathrm{d}M_r/\mathrm{d}r$, $\mathrm{d}P/\mathrm{d}r$, $\mathrm{d}L_r/\mathrm{d}r$, and $L_r = f(\mathrm{d}T/\mathrm{d}r)$], and to review the five constitutive relations:

(1) The perfect gas law:

$$P_{\text{gas}} = \frac{k}{m_{\text{H}}} \frac{\varrho}{\mu} T,$$

(2) The law governing the radiation pressure:

$$P_{\text{rad}} = \tfrac{1}{3} a T^4,$$

(3) The expression for the total pressure P:

$$P = P_{\text{gas}} + P_{\text{rad}},$$

(4) The Kramers-Rosseland opacity law:

$$\bar{k} = k_0 \varrho^{2-\alpha} T^{-3.5}$$

(where α is often taken to be zero, but where we have used $\alpha = 0.25$),

(5) The law of nuclear energy generation, which in its 'empirical' form is written:

$$\varepsilon = \varepsilon_{01} \varrho T^{\nu_1} + \varepsilon_{02} \varrho T^{\nu_2} + \cdots.$$

(We recall that μ, k_0, ε_{01}, ε_{02}, etc., depend on X, Y, Z (X_{CN}), etc.) This enabled us to show, taking as an example the Sun (in its present state of evolution), that it is possible to satisfy all the differential and constitutive relations simultaneously for a certain distribution $\varrho(r)$, $X(r)$. Thus we showed that a problem of this type can have *at least one* solution (in spite of the numerous relations to be satisfied) and that this solution is probably *unique* because of the *physical* requirement imposed on the parameters: that of satisfying the differential relations, the boundary conditions, and the constitutive relations at the same time.

We must still specify the *mathematical structure* of the problem and, without entering into the details of certain 'technical tricks', explain the *basic principles* governing the *actual* determination of the distributions $\varrho(r)$ and $X(r)$ for each model.

1.1. The Advantage of Studying 'Evolutionary Sequences'

The distribution $X(r)$ introduces an important physical concept which we have not yet discussed: we must not isolate the different 'evolutionary states' of a star in an individual manner, but rather we must follow '*step by step*', *beginning with a* '*prestellar*' *state* (of homogeneous chemical composition $X=X_0$, $Y=Y_0$, etc.), the evolution of a gaseous sphere of *constant* mass M_0.

In the 'initial' state this mass has just finished its (gravitational) *contraction* from the 'interstellar material' and is just beginning to 'burn' its principal 'nuclear fuel': hydrogen.

1.2. The 'Fossilized' Composition

The determination of $\varrho(r)$, $P(r)$, and $T(r)$ in these chemically homogeneous models obviously does not require a knowledge of the function $X(r)$ (as we have already seen for the Sun, for r' greater than 0.30). Even so, one must know the *initial composition* (X_0, Y_0, Z_0) of the star.

Now, a fortunate circumstance enables us to avoid choosing (X_0, Y_0, Z_0) *arbitrarily*, at least when we want to know the structure of a real star which is *actually observable*, but which is *observed in an advanced evolutionary state*. In this state the nuclear reactions have already modified (X_0, Y_0, Z_0) and created variations of X, Y, and Z with r – that is, inhomogeneities in the composition: $X(r)$, $Y(r)$, and $Z(r)$ [which must not be confused with *inhomogeneities in the density* $\varrho(r)$, *which are present even in a star of homogeneous composition* and which are also associated with a variation of $T(r)$].

But it can be shown (see Note 1 below) that neither *rotation* (with the *meridian circulation* it provokes) nor *convection* can normally produce a '*general mixing*' – that is, bring material from the 'core' (where nuclear reactions occur) into the external *envelope*, where it is too cold for ε_r to be different from zero. [*N.B.* – This does not, however, exclude the possibility that the core *or* the envelope, independently, can be the site of *convective* motions due to a temperature gradient leading to a mechanical instability.]

Now, the *absence* of 'general mixing' means that the 'external' composition (that of the outer envelope) of the (*present*) observable 'evolved' phase – which can be determined by quantitative spectral analysis – is none other than the '*fossilized*' (prehistoric) composition of the *initial phase*.

In other words, (X_0, Y_0, Z_0) and X_{CN} as *now* observed at the surface are the same as (X_0, Y_0, Z_0) of the '*initial*' *chemically homogeneous* star, with which we must *begin* the study of 'evolutionary sequences'.

Evolution causes the radius R and the luminosity L of the star to change (very slowly) with time, whereas the mass M can be considered constant, in spite of the mass lost in the form of radiation or through the 'escape' or 'ejection' of particles (see Note 2 below).

In any case, there is no factor that can influence either the mass M or the conformity of the envelope composition with the initial homogeneous composition in the *initial phases of evolution* (approximately 10^8 years) of normal main-sequence stars. It is these stars that Schwarzschild studies in his remarkable treatise [1].

Note 1. On Stellar Rotation

Rotation can provoke a general mixing by *meridian circulation*. But this circulation is a function of the *speed* of rotation. Now, for most main-sequence stars of *type F or later* (and thus in particular the Sun) – that is, for the vast majority of stars – the rotational speeds observed (at the present time!) are so small that they can produce only a negligible amount of 'general mixing'.

For stars of type A and B ('early main-sequence stars'), whose surface temperatures T_s are greater than or equal to about 10000 K, the observed *rotational velocities* W_0 are (even now!) *very large*. Thus, if we follow the custom of expressing the *angular* velocities Ω in terms of linear equatorial velocities $W_0 = \Omega R$, we have the following approximate values for stars of various spectral types Sp:

Sp	T_s	W_0
B	25000 K	210 km/sec
A	10000 K	170 km/sec
F	8000 K	30 km/sec

Nevertheless, a more detailed study shows that for *normal* stars of this type the effects of rotation on general mixing remain negligible. This does not exclude extreme cases of high-speed rotation (or of close binaries such as the W Ursae Majoris stars, where the circulation is initiated by tidal effects), which remain difficult to study.

Note 2. Processes Capable of Modifying the Mass M

It is easy to prove that even during intervals Δt of the order of 10^9 years, the luminosity $L_\odot$ of the Sun, for example, leads to energy losses

$$\Delta t \cdot L_\odot = 10^9 \times (3 \times 10^7) \times (4 \times 10^{33}) = 12 \times 10^{49} \text{ ergs},$$

equivalent to a mass loss of the order of:

$$\Delta M_\odot = 12 \times 10^{49}/c^2 = 12 \times 10^{49}/9 \times 10^{20} \approx 10^{29} \text{ grams}.$$

Such a loss represents only $\Delta M_\odot / M_\odot = 1/20000$ of the mass of the Sun.

Thus it would seem that for stars such as the Sun, the mass lost by radiation is practically negligible for a period of one billion years.

In reality, however, the problem of the 'conservation' of mass is more complex. All stars, especially 'hot' stars, have a luminosity L_S which increases much faster than their mass M_S. We have approximately

$$L_S = L_\odot (M_S/M_\odot)^5. \tag{1}$$

If we put

$$M = M_S/M_\odot \quad \text{and} \quad L = L_S/L_\odot\,, \tag{2}$$

Equation (1) can be written:

$$L = M^5\,. \tag{3}$$

And we will have:

$$\frac{\Delta M_S}{M_S} = \frac{(L_S \cdot \Delta t)\ \text{ergs}}{(M_S \cdot c^2)\ \text{ergs}} = \frac{L\,(L_\odot\ \Delta t)}{M\,(M_\odot c^2)} = 5 \times 10^{-5}\,M^4\,. \tag{4}$$

Thus we find that for $M = 10$ – that is, for a star 10 times more massive than the Sun – the relative change of the mass in 10^9 years can be as great as $\frac{1}{2}$, which is not at all negligible.

In addition, the star can lose mass by the 'escape' of particles ('evaporation') due to thermal motion (stimulated by P_{rad}). The star can also lose mass by 'ejection' due to electromagnetic processes (solar wind). Finally, the star can *gain* mass by '*accretion*', i.e. *by sweeping up the interstellar gas as it moves*. But, in fact, the velocity of the stars with respect to the 'interstellar clouds' is *too great* (at the present time!) for notable *accretion* (once again, we encounter the paradoxical idea of the efficiency of interactions between particles or objects whose relative motion is *slow*).

On the whole, all these phenomena have an important effect only on the *red giants*.

Note 3. *The Fossilized Helium*

It may be surprising that X_0 at the beginning of evolution is different from 1, which implies that helium is present *before* the conversion of hydrogen into helium by thermonuclear reactions in the stellar core.

This is due to the fact that the 'beginning' of star formation (from the diffuse gas) occurs later than the 'initial' phase of the expansion of the Universe. During this very hot and very dense phase, a certain amount of helium could have been formed. This explains why one finds helium at the *surface* of very old stars (in the absence of 'general mixing').

This question is presently the object of numerous theoretical and observational investigations, and the problem cannot be considered entirely solved.

2. The Evolution of the Distributions $X(r)$ and $Y(r)$

After the 'initial' model whose *homogenous* composition X_0, Y_0, Z_0 is *known*, Schwarzschild considers a *discrete* series of stellar models $\{S\}_t$. These models correspond to a time step Δt small enough so that the composition $X(r)_{t+\Delta t}$, $Y(r)_{t+\Delta t}$, etc., at each 'point' r of the model $\{S\}_{t+\Delta t}$ can be determined from that of the 'preceding' model $\{S\}_t$ by the differential relationship

$$X(r)_{t+\Delta t} - X(r)_t = \left[\frac{\partial X(r)}{\partial t}\right]_t \Delta t\,, \tag{5}$$

and by analogous equations for $Y(r)$ and $Z(r)$.

In addition to the variation of composition in each element of the star where nuclear reactions take place, evolution is generally accompanied by an *expansion* or *contraction*. Thus one cannot use the variable r' or even r in equations such as (5). This difficulty is surmounted by a physical-mathematical artifice which consists of taking M_r instead of r as the variable that determines the position of an element. This is equivalent to defining a sphere (Σ_r) by the mass of the matter it encloses, rather than by its radius. Equations (5) are then replaced by equations such as:

$$X(M_r)_{t+\Delta t} - X(M_r)_t = \left[\frac{\partial X(M_r)}{\partial t}\right]_t \Delta t\,. \tag{5'}$$

The advantage of this change of variables is obvious. Since the transmutations transform protons into alpha particles, or produce other nuclear transformations with 'conservation' of mass (except for the binding energies), the mass M_r of *a given set of particles* remains constant in time, even when r varies because of contraction or expansion or when the particles themselves change. When the functions $\varrho(M_r)$, $P(M_r)$, $T(M_r)$, $L_r(M_r)$ and especially $r(M_r)$ have been found, a purely mathematical 'inversion' makes it possible to return to the variable r.

In other words, we shall use the variable M_r to connect two successive evolutionary stages. But for the discussion of each stage, we can use the variable r, which is physically clearer.

We note in particular that the functions $X(M_r)$, $Y(M_r)$, and $Z(M_r)$ in a given evolutionary stage are determined not only by the usual equilibrium equations (mechanical, energy, etc.), but also by the results of the nuclear transmutations that took place *during the preceding evolutionary phases*. This makes it necessary to follow *all* these phases, from a homogeneous mixture of given composition up to the phase under consideration.

Let us now return to Equation (5′), and to the analogous equations for $Y(M_r)$ and $Z(M_r)$. The coefficient of Δt on the right-hand side is *completely known* (if one has used the prescribed method, i.e. examined all the intermediate phases between $t=0$ and t). Actually, the *variation* $\Delta X(M_r)$ of $X(M_r)$ during the time Δt inside a layer $(M_r, \mathrm{d}M_r)$ located between spheres of mass M_r and $(M_r+\mathrm{d}M_r)$ is proportional to the energy $[\varepsilon(M_r)]_t$ produced per gram per second near the time t in the corresponding layer; the constant of proportionality is easily obtained by the following argument.

In the p-p chain, for example, the energy Q_{pp} liberated during the formation of each helium *atom* (I emphasize *atom* and not nucleus, in order to take into account the mass of the negative electrons destroyed or recovered) by the transmutation of four hydrogen *atoms* – that is, by using up a mass of $4m_{\mathrm{H}}$ grams of hydrogen (where m_{H} is the inverse of Avogadro's number) – is

$$Q_{\mathrm{pp}} = 26.2\ \mathrm{MeV} = 42.1 \times 10^{-6}\ \mathrm{ergs}\,. \tag{6}$$

Thus if we follow Schwarzschild's notation and let $\varepsilon^*_{\mathrm{pp}}$ be the energy in ergs supplied by each gram of hydrogen consumed in the p-p chain, we will have:

$$\varepsilon_{\text{pp}}^{*} = \frac{Q_{\text{pp}}\,\text{ergs}}{(4m_{\text{H}})\,\text{grams}} = \frac{Q_{\text{pp}}\,\text{ergs} \times N_A}{4} = 6.34 \times 10^{18}\,. \tag{7}$$

And by an analogous argument, we find

$$\varepsilon_{\text{CN}}^{*} = 6.0 \times 10^{18}\ \text{ergs} \tag{7'}$$

for each gram of hydrogen consumed in the C-N cycle.

Now we know ε_{pp} and ε_{CN} (without the asterisk), i.e. the energies supplied per second by each gram of the stellar mixture. But in each gram of this mixture in the layer $(M_r, \mathrm{d}M_r)$ there are, at the time t, $X(M_r)$ grams of hydrogen and $Y(M_r)$ grams of helium; and the (negative) variation $\Delta X(M_r)$ of $X(M_r)$ during Δt seconds is the sum of the variation $\Delta_{\text{pp}}X(M_r)_t$ and the variation $\Delta_{\text{CN}}X(M_r)_t$, which are given by the relations

$$(\varepsilon_{\text{pp}}^{*})\cdot(\Delta_{\text{pp}}X) = -\,\varepsilon_{\text{pp}}\cdot\Delta t \quad \text{and} \quad (\varepsilon_{\text{CN}}^{*})\cdot(\Delta_{\text{CN}}X) = -\,\varepsilon_{\text{CN}}\cdot\Delta t\,, \tag{8}$$

respectively. These express, in two equivalent manners, the energy released during the time Δt by the p-p and C-N processes for each gram of the mixture $(M_r, \mathrm{d}M_r)$.

It then follows that we have, in an obvious notation:

$$\begin{aligned} \Delta_{\text{pp}}Y &= (-\,\Delta_{\text{pp}}X) = \left[\frac{\varepsilon_{\text{pp}}}{\varepsilon_{\text{pp}}^{*}}\right]\cdot\Delta t \\ \Delta_{\text{CN}}Y &= (-\,\Delta_{\text{CN}}X) = \left[\frac{\varepsilon_{\text{CN}}}{\varepsilon_{\text{CN}}^{*}}\right]\cdot\Delta t\,. \end{aligned} \tag{9}$$

The total variation ΔX during the time Δt is given by $\Delta_{\text{pp}}X + \Delta_{\text{CN}}X$, so that we have:

$$\frac{\partial X\,(M_r)}{\partial t} = -\left(\frac{\varepsilon_{\text{pp}}}{\varepsilon_{\text{pp}}^{*}} + \frac{\varepsilon_{\text{CN}}}{\varepsilon_{\text{CN}}^{*}}\right). \tag{10}$$

For temperatures greater than $T_6 \approx 100$, a small negative variation $\Delta_{3\alpha}Y$, due to the conversion of three α particles $(=3\text{He}^4)$ into a C^{12} nucleus, must be added to the positive variation $(\Delta_{\text{pp}}Y + \Delta_{\text{CN}}Y)$ of Y. Thus in an obvious notation, we have:

$$\frac{\partial Y\,(M_r)}{\partial t} = +\left(\frac{\varepsilon_{\text{pp}}}{\varepsilon_{\text{pp}}^{*}} + \frac{\varepsilon_{\text{CN}}}{\varepsilon_{\text{CN}}^{*}}\right) - \frac{\varepsilon_{3\alpha}}{\varepsilon_{3\alpha}^{*}}\,. \tag{11}$$

3. Discussion

We have already mentioned, in Section 6 of Chapter III, that a gaseous layer becomes *unstable* (convective motions) when the *absolute value of the (relative) temperature gradient exceeds* a certain fraction (0.4) of the absolute value of the (relative) pressure gradient. We have also seen that *partial* ionization favors convection. The principal effect of convection which interests us here is that it produces a partial mixing which tends to make the layer chemically *homogeneous*.

Now, when we consider, for example, three stellar models having masses of 10, 2.5, and 1.0 solar masses, we find that their *central* temperatures T_0 (center) in the *initial* homogeneous phase are given by Table XX below. (We should not be surprised to find that the central temperature for $M = 1.0$ is not the same as we found above for the Sun, in a phase which *was not* the initial phase.)

This difference in the central temperature in the initial phase (it persists in later phases) results in *two different types of structure*, depending on whether M is *larger* or smaller than 1.7. The former corresponds to the '*upper* main sequence' and the latter to the '*lower* main sequence', following the mass distribution of main-sequence stars (the only stars we shall consider here). In this section we shall refer to stars whose mass M is between 1.7 and 11 as 'massive' stars, and to stars whose mass is between 1.7 and 0.8 as 'low-mass' stars. A detailed analysis (cf. B. Strömgren, in [4], pp. 269–95, and especially pp. 291–92) shows that among stars whose mass is smaller than 1.7, those whose mass is smaller than 0.8 have certain peculiarities which oblige one to study them separately.

TABLE XX

$M/M_\odot$	10	2.5	1.0
T_0 (center)	28×10^6	20×10^6	12×10^6

Of course, all parameters – and in particular, the central temperature in the initial phase (given by the above table) – depend for any given mass on the *initial chemical composition* X_0, Y_0, Z_0. Since the sum $(X_0 + Y_0 + Z_0)$ is always equal to 1, it suffices to give *two* of these three quantities. It is easier to distinguish differences of composition in X_0 and Z_0 than in X_0 and Y_0, since Z_0 is generally a number with few significant figures.

It is found, however (cf. Strömgren, *loc. cit.*) that for 'massive' stars the influence of (X_0, Z_0) for reasonable values of X_0 and Z_0 (X_0 from 0.60 to 0.90 and Z_0 from 0.01 to 0.04) remains very small. This is why we neglected to specify in Table XX (giving the central temperatures in the initial phase) that the composition for the first two models was $X_0 = 0.90$ and $Z_0 = 0.01$, whereas the composition of the last model (which is not 'massive') was $X_0 = 0.744$ and $Z_0 = 0.02$.

Table XXI, derived from Strömgren's tables, shows the relative smallness of variations in the luminosity L and the radius R for a 'zero-age' (initial phase) 'massive' star for the two most extreme cases (of different composition) considered by Strömgren. (We are interested mainly in the influence on L; for the influence on R is very small in any case.)

TABLE XXI

Mass	$X_0 = 0.60$, $Z_0 = 0.02$		$X_0 = 0.70$, $Z_0 = 0.03$	
1.78	$L = 21.4$	$R = 1.4$	$L = 11.3$	$R = 1.5$
2.82	$L = 123$	$R = 1.9$	$L = 65$	$R = 2.0$
4.47	$L = 646$	$R = 2.5$	$L = 371$	$R = 2.6$
7.08	$L = 3090$	$R = 3.2$	$L = 1950$	$R = 3.3$

a. '*Massive*' *Stars*

The essential characteristic of 'massive' main-sequence stars (M greater than 1.7) is the existence of a '*convective core*'; that is, the region where thermonuclear reactions take place (and where ε is not zero, and $L' = L_r/L$ is different from 1) is the site of convective motions. In such a 'core' the relation between dT/dr and the opacity $\bar{k}$ is not applicable; but in return, we have the simple relation (already encountered in Chapter III, Section 6; see also Chapter III, Section 5):

$$n = \text{'polytropic index'} = (d\varrho/\varrho)/(dT/T) = 1.5\,;$$

or the equivalent relations

$$n + 1 = \frac{d \log P}{d \log T} = 2.5 \qquad \frac{1}{T}\frac{dT}{dr} = (0.4)\frac{1}{P}\frac{dP}{dr}.$$

It so happens that the condition for 'stability' (the absence of convection) and thus for the exchange of energy by radiation alone, is that $(n+1)$ should be *greater* than 2.5.

But this condition *is not satisfied* (in 'massive' stars, with distributions $\varrho(r')$ and $X_0 = \text{const}$ that satisfy all the conditions of the problem) for values of r' *smaller* than a certain 'limiting' value (the boundary of the convective core). Table XXII gives some details concerning the 'convective core' of four models, after Strömgren and Schwarzschild. We note, among other things, that regardless of the composition, the convective core is *relatively larger* (in r') when the mass is larger. In this case it also contains a larger fraction M' of the total mass M. We also note, paradoxically, that ϱ_c at the center decreases when the mass M increases.

TABLE XXII

Composition	$X_0 = 0.70,\ Z_0 = 0.03$		$X_0 = 0.90,\ Z_0 = 0.01$	
Mass M	2.82	7.08	2.5	10.0
Boundary r'_{lim} of C.C.[a]	0.15	0.21	0.16	0.24
$M'(r'_{\text{lim}})$	0.18	0.27	0.18	0.26
$T_6(r'_{\text{lim}})$	16.5	20.0	15	19
$\varrho(r'_{\text{lim}})$	24	6.6	31	4.4
T_6 (center)	23	28	20	28
ϱ_c (g/cm^3)	38	12	48	8
R	1.96	3.3	1.59	3.63
$\bar{\varrho}$ (g/cm^3)	0.53	0.28	0.88	0.30

[a] C.C. = 'convective core'.

The last line of Table XXII explains why the central density ϱ_c decreases as the mass increases: the *mean* density ϱ in g/cm^3 for the stars under consideration *decreases* as the mass increases, because the radius R appears to the third power in the formula for $\bar{\varrho}$. Thus we might expect ϱ_c also to decrease as the mass increases.

We could try to give a physical explanation for the existence of convection in the core of 'massive' stars by noting the fact that the central temperature of 'homologous' stars (with a universal function for $\varrho(r')/\bar{\varrho}$) varies as M/R. At least, this is what we find 'empirically' in Table XXIII, which is derived from Table XXII.

The quantity $[T_6\,(\text{center})/(M/R)]$ is constant only to within an order of magnitude. But this is enough to give central temperatures T_6 between 20 and 28. Thus we are in the domain of the C-N cycle, and the energy output depends on a high power ($v=17$) of the temperature.

TABLE XXIII

Composition	$X_0=0.70,\ Z_0=0.03$		$X_0=0.90,\ Z_0=0.01$	
Mass M	2.82	7.08	2.5	10.0
Mass M/Radius R	1.44	2.14	1.57	2.75
T_6 (center)/(M/R)	16.0	13.0	12.7	10.2

It would seem that this high value of v is responsible for the existence of the convective core. But the relationship between these two phenomena *is not at all direct*, as is easily seen when one tries to show the influence of v on the ratio

$$\mathrm{d}\,(\log P)/\mathrm{d}\,(\log T),$$

consulting the work of Cowling [7] or those of Wrubel [5, p. 61], and Tayler [3].

Of course, the convection in the core maintains during evolution a chemical composition $X(\text{core})_t$, $Y(\text{core})_t$, etc., which varies with time but always remains 'homogeneous' (i.e., independent of r or of M_r). On the other hand, the region outside the core also maintains a 'homogeneous' composition – the initial composition X_0, Y_0, Z_0. There results a *discontinuity* in composition on the sphere of radius r'_{lim}, but this does not imply a discontinuity in ϱ, T, P, or L_r.

b. *Low-Mass Stars*

The principal characteristic of these stars in the initial phase (but also in the Sun, even in an 'advanced' phase of evolution) is the presence (in addition to a perfectly 'normal' core in radiative equilibrium) of an '*outer*' *convective region*, whose thickness is generally non-negligible.

In this case, the (radiative) stability of the core appears to be associated with the low central temperature of these stars, implying energy production by the p-p chain, which depends on a relatively low power of T.

The existence of the outer convective region is explained by the partial ionization of hydrogen in this zone, as we have seen in the case of the Sun.

The constant in the equation $T=\text{const}\times P^{0.4}$, which in a convective region replaces the equation connecting $\mathrm{d}T/\mathrm{d}r$ with the opacity, is very difficult to determine for an *outer* convective zone; for in the observable region, T and P both go to zero simultaneously. This constant must be considered as an additional unknown, which one tries to determine from a study of the stellar *atmosphere*.

As an example of the 'initial state' of a 'low-mass' star, we can take the initial state of the Sun, and compare it with the 'evolved' state ($X_c=0.50$), which we have studied in detail. We have, then, the following data for the Sun [1, Table 28.3]:

TABLE XXIV

'Initial' state of the Sun	Present state of the Sun
$X_0=0.744$ $Z_0=0.02$	$X_0=0.744$ $Z_0=0.02$
$M_\odot=1$	$M_\odot=1$
$r'_z=0.88$	$r'_z=0.86$
$M'(r'_z)=0.999$	$M'(r'_z)=0.998$
$T_6(r'_z)=0.72$	$T_6(r'_z)=0.88$
T_6 (center) $=12.39$	T_6 (center) $=14.6$
$R_0=1.021$	$R_\odot=1$
$\bar{\varrho}=1.32$ g/cm³	$\bar{\varrho}=1.41$ g/cm³
$\varrho_c=77$ g/cm³	$\varrho_c=135$ g/cm³
$L_0=0.578$	$L=1$
$X_c=X_0=0.744$	$X_c=0.50$

As we see, the principal changes produced by an evolution which decreases X(center) to 0.50 are: a slight increase in the central temperature, which goes from $T_6=12.39$ to $T_6=14.6$; a slight decrease (hardly noticeable) in the radius R from $R_0=1.021$ to $R=1$; a slight increase in the mean density, which goes from $\bar{\varrho}=1.32$ to $\bar{\varrho}=1.41$; a relatively large increase in the central density, which goes from 77 g/cm^3 to 135 g/cm^3 (a change essentially due to a decrease in X_c, which increases μ and the central pressure P_c, whose common logarithm goes from 17.13 to 17.35). But the most spectacular change (in contradiction to an oversimplified 'mass-luminosity relation') is in the luminosity L, which goes from

$$L_0 = 0.578 \quad \text{to} \quad L = 1\,.$$

As for the relative position of the convective zone defined by r'_z, it remains practically unaffected by the evolution; for $r'_z=0.88$ is a value very close to that already found for the Sun in its present state.

4. The Mathematical Structure of the Problem. Principles of the Integration Methods

Although *physically* the function $\varrho(r)$ is, so to speak, the keystone of the internal structure of a star, it is advantageous from the *mathematical* point of view to eliminate it as soon as possible; for it is the only one of the important unknown functions which does not appear as a derivative in any of the fundamental relations. Moreover, the variable r can be retained for each evolutionary stage.

If, in order to shorten the presentation, we neglect P_{rad} in comparison with P_{gas}, we will have the perfect gas law in the form:

$$\varrho = \frac{\mu m_{\text{H}}}{k}\frac{P}{T} = A_\mu \frac{P}{T}. \tag{12}$$

Then our four fundamental differential equations (written in the right-hand column below) appear in the following form (with* $\bar{k}=k_0\varrho^2T^{-3.5}$ and $\varepsilon=\varepsilon_0\varrho T^\nu$):

$$\frac{dM_r}{dr} = 4\pi A_\mu\left(\frac{P}{T}r^2\right) \qquad (13) \qquad \frac{dM_r}{dr} = 4\pi r^2\varrho$$

$$\frac{dP}{dr} = -GA_\mu\left(\frac{P}{T}\frac{M_r}{r^2}\right) \qquad (14) \qquad \frac{dP}{dr} = -\frac{GM_r}{r^2}\varrho$$

$$\frac{dT}{dr} = -B_\mu\left(\frac{L_rP^2}{r^2T^{6.5}}\right) \qquad (15) \qquad \frac{dT}{dr} = -\frac{3k_0L_r\varrho^2}{16\pi acr^2T^{6.5}}$$

$$\{T=\text{const}\times P^{0.4} \quad \text{if there is convection}\}$$

$$dL_r/dr = C_\mu(r^2P^2T^{\nu-2}) \quad (16) \qquad dL_r/dr = 4\pi r^2\varepsilon\varrho = 4\pi r^2\varepsilon_0\varrho^2T^\nu .$$

In these equations,

$$A_\mu = \mu m_H/k \qquad B_\mu = 3k_0A_\mu^2/16\pi ac \qquad C_\mu = 4\pi\varepsilon_0A_\mu^2 .$$

Of course, in the 'initial phase' A_μ, B_μ, and C_μ depend on X_0, Y_0, and Z_0 through μ, k_0, and ε_0; and in subsequent phases they depend on X, Y, and Z. In the equations in the left-hand column we have collected within parentheses the variable r and the quantities $P(r)$, $T(r)$, $M_r(r)$, and $L_r(r)$, which are functions of r. As we see, from the mathematical point of view we are now faced with a system of four differential *equations* (13) to (16) with four *unknown functions* of the independent variable r: $M_r(r)$, $P(r)$, $L_r(r)$, and $T(r)$.

If we indicate by the subscript zero the fact that we are originally concerned with the 'initial phase' of evolution, with homogeneous chemical composition, and attach the subscript c to quantities that refer to the *center*, we have the following boundary conditions:

$$\begin{array}{lllll} \text{Surface: } r = R_0 & M_r = M_0 & P = 0 & L_r = L_0 & T = 0 \\ \text{Center: } r = 0 & M_r = 0 & P = P_c^0 & L_r = 0 & T = T_c^0 . \end{array} \qquad (17)$$

When we study an evolutionary sequence beginning with an initial state whose composition X_0, Y_0, Z_0 is *assumed* (in order to construct a series of models, among which we can later choose by comparing the results with the observations), the parameters μ, k_0, and ε_0 – and consequently, the parameters A_μ, B_μ, and C_μ – are known constants. We also assume the mass M_0 of the model we want to construct (this is the *observed* mass of a star whose structure is to be found).

On the other hand, we do not know at the outset the intial value R_0 of the radius of the sphere occupied by M_0, nor the initial luminosity L_0, nor the initial central pressure P_c^0, nor the initial central temperature T_c^0. (As a warning against an easily committed error, we note in particular that R_0 for the Sun is not *a priori* equal to the present value

* We recall that for the Sun ([9], Chapter VII) we have $k_0 = 2.4\times10^{23}(1+X)^{0.75}$ and in the expression for $\bar{k}$, ϱ^2 is replaced by $\varrho^{1.75}$.

of the solar radius $R_\odot$. It happens that for $X_0 = 0.744$ and $Z_0 = 0.02$, the values used by Schwarzschild,

$$R_0/R_\odot = 1.021$$

that is, approximately 1. But for the same composition, $L_0/L_\odot$ is equal to only 0.578.)

Nevertheless we can, in principle, begin all our integrations at the surface by choosing the value of R_0 *arbitrarily*. Let us also choose the value L_0 of the function L_r at the point $r = R_0$ (in the initial phase) *arbitrarily*. Thus we will begin with two arbitrary parameters R_0 and L_0; and we can construct (Figure 15) the points S_M, S_P, S_L and S_T on the curves representing $M_r(r)$, $P(r)$, $L_r(r)$, and $T(r)$ at the point $r = R_0$ – that is, at the surface S of the star; for we will have the common abscissa $r = R_0$ of these points, as well as the corresponding ordinates M_0, 0, L_0, 0.

Then, using Equations (13), (14), (16), and (15), which give the derivatives dM_r/dr, dP/dr, dL_r/dr, and dT/dr in terms of $M_r = M_0$, $P(0) = 0$, $L_r(0) = L_0$, and $T(0) = 0$, we can in principle compute the values of the derivatives in question at the points S_M, S_P, S_L, and S_T. Thus we will have the *slopes of the tangents to the curves* representing our

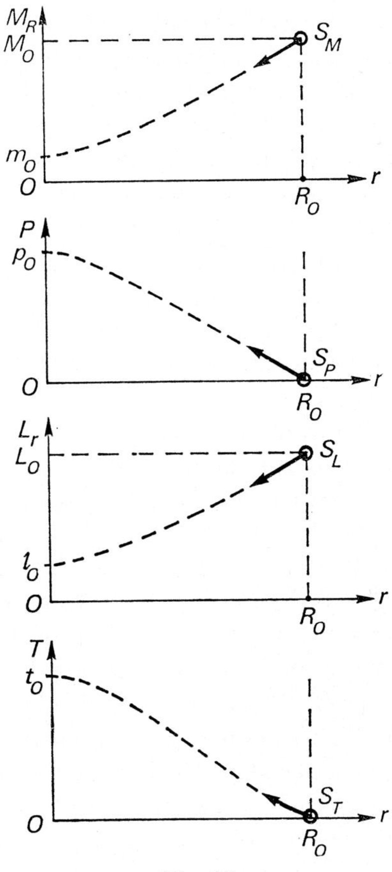

Fig. 15.

four unknown functions, at the surface S. This is indicated in Figure 15 by the arrows originating at the points S_M, S_P, S_L, and S_T; in this figure we have deliberately ignored, in the interest of clarity, the fact that $M_r(r)$ remains very close to M_0 and that $L_r(r)$ remains very close to L_0 when r is close to R_0.

Continuing in this way, step by step – that is, by applying the classical methods for the numerical integration of systems of differential equations – we can follow the graph of our four curves towards $r=0$, all the way to the center. But in general, if R_0 and L_0 are truly arbitrary, the final situation will be as indicated by the dashed curves in Figure 15. Thus for $r=0$, the curve $M_r(r)$ will pass through a point m_0 whose ordinate $M_r(0)$ is different from zero, and the curve $L_r(r)$ will pass through a point l_0 whose ordinate $L_r(0)$ is different from zero.

Now, we seek solutions for which $M_r(0)=0$ and $L_r(0)=0$, while we have *no preconceived idea* concerning P_c^0 (the central value of the initial pressure) and T_0 (the central value of the initial temperature); thus the position of the points p_0 and t_0 on the curves representing $P(r)$ and $T(r)$ is *indifferent* to us, for the time being.

At this point, it is obvious that a methodical search for the solution could be reduced to the construction of a grid of curves like those in the figure, for a 'rectangular array' of values of R_0 and L_0 which are chosen arbitrarily but in sufficient number, and with a reasonable order of magnitude. An interpolation between the values of R_0 and L_0 which give m_0 and l_0 *near* the origin then gives the pair (R_0^S, L_0^S) corresponding to the 'correct solution' – the one which makes the two curves, the one representing M_r and the one representing L_r, pass through the origin of the coordinate system.

If we wish to conduct this search systematically, we begin by assuming a set of values R_0^1, R_0^2, R_0^3, etc., of R_0. For *each of them* we can, by varying L_0, find that value L_0^n which, combined with a particular value R_0^n, makes the curve $M_r(r)$ pass through the origin – without, however, having the corresponding curve $L_r(r)$ necessarily pass through the origin as well. Only one particular value R_0^S of R_0^n will make $L_r(r)$ pass through the origin, and the corresponding pair (R_0^n, L_0^n) will give the desired solution by interpolation.

The curves representing $P(r)$ and $T(r)$ which correspond to R_0^S and L_0^S will give the exact values, hitherto unknown, of P_0^c and T_0^c corresponding respectively to the ordinates of p_0 and t_0 on the 'right curves'. Then the perfect gas law (12) will give the 'right' distribution $\varrho(r)$.

Remarks. In general, the method we have just described 'converges' poorly when it is applied beginning at $r=R_0$, where $P=0$ and $T=0$. This is especially true for the function $T(r)$, for T appears with a very large exponent in the denominator of the expression (15) for $\mathrm{d}T/\mathrm{d}r$. We surmount this difficulty by looking for an *analytical* expression for the limiting value of $P^2/T^{8.5}$ in the neighbourhood of $r=R_0$.

Moreover, the 'classical' form of Equation (14) (written in the right-hand column) shows that in the neighbourhood of $r=0$, where ϱ remains *finite* and takes on the value $\varrho_c=\varrho(0)$, we can consider (replacing M_r by $\frac{4}{3}\pi r^3\varrho_c$) that $\mathrm{d}P/\mathrm{d}r$ varies as $r^3/r^2=r$, and consequently tends to *zero* when $r\to 0$. However, numerical computation shows that $\mathrm{d}P/\mathrm{d}r$ passes through a very sharp maximum before going to zero for $r=0$. It is therefore difficult to determine P_0 precisely by the method which consists of starting from $r=R_0$ and approaching the 'difficult' region near $r=0$ with a certain accumulation of numerical errors, making the large value of $\mathrm{d}P/\mathrm{d}r$ especially uncertain.

This is why, instead of taking R_0 and L_0 as the only free parameters, it is preferable to take P_c and T_c as *additional* provisional parameters; then, starting from $r=0$, we construct (Figure 16) curves (C_{center}) for each of the four unknown functions, for which all the ordinates are now known from the outset.

These new curves (C_{center}) do not necessarily join the curves (C_{surface}) which have been previously discussed, and which can be drawn starting from $r=R_0$ once R_0 and L_0 have been assumed.

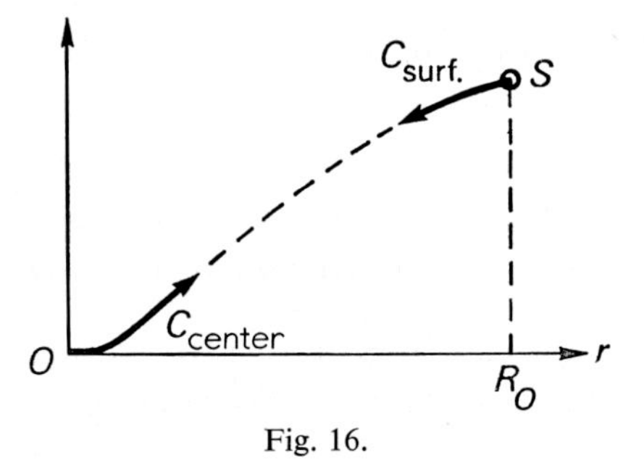

Fig. 16.

Since the curves (C_{center}) originating at the center already satisfy the conditions $M_r(0)=0$ and $L_r(0)=0$, the test for the 'right' values of R_0, L_0, T_0 and P_0 will now be to make the four curves (C_{center}) join the corresponding curves (C_{surface}) at some point of abscissa $r=r_1$.

5. The Age of a Star

For more advanced stages of evolution, we can apply the methods just described, interchanging the roles played by r and M_r.

Now let t be the *sum* of the (Δt) corresponding to the (ΔX), (ΔY), and (ΔZ) which give a solution R_t and L_t, where R_t and L_t represent the values observed *at the present time* for certain stars (for very distant stars, one must of course correct for the travel time of the light). Then $t=\sum(\Delta t)$ represents the 'age' of the star – that is, the interval separating its present state from its 'initial' state. In this way, we find that the present radius and luminosity of the Sun are reached (starting with $X_0=0.744$ and $Z_0=0.02$) in $t=$approximately 5×10^9 years, with $X_t(\text{center})=$approximately 0.50.

Table XXV – which we have taken from Strömgren ([4], p. 278, Table 4) – enables us to follow in detail, as a function of the values $X_t^{(c)}$ at the center, the accumulation of Δt leading to various ages t_6 in millions of years ($t_6=t/10^6$) for stars of mass $M=2.82$ ($\log M=0.45$) and $M=7.08$ ($\log M=0.85$) with initial composition $X_0=0.70$, $Z_0=0.03$.

In this table we note that stars of the type under consideration undergo an *expansion* as they age, while the *central temperature increases* (relatively slowly). The evolution of stars of this type is more *rapid* if they are more *massive*, since the state in which

TABLE XXV

Mass = 2.82					Mass = 7.08				
$X_t^{(c)}$	t_6	R_t	$T^t{}_{6,c}$	L_t	$X_t^{(c)}$	t_6	R_t	$T^t{}_{6,c}$	L_t
0.70	0	1.96	23	65	0.70	0	3.3	28	1950
0.60	70	2.11	23	71	0.60	9	3.6	29	2089
0.50	120	2.28	24	78	0.50	15	3.9	29	2344
0.40	170	2.46	24	85	0.40	21	4.3	30	2570
0.30	210	2.67	25	85	0.30	26	4.7	31	2818
0.20	240	2.91	25	85	0.20	29	5.2	32	2818
0.10	260	3.15	27	85	0.10	32	5.8	33	3090

$X=0.10$ at the center is reached in 260 million years for $M=2.82$, but in only 32 million years for $M=7.08$.

6. The Relations between P, T, L, R, and Parameters such as M, k_0, ε_0, and μ for 'Homologous' Models. The 'Mass-Luminosity' and 'Mass-Radius' Relations

We can now return to the calculations performed on 'homologous models' in Chapter II (Section 7), and carry them a little further. We assumed that the function

$$D(r') = \varrho(r')/\bar{\varrho} \tag{18}$$

is a 'universal' function, applicable to all stars.

Naturally, this hypothesis has a better chance of being correct if, instead of applying it to *all* stars, we limit ourselves to comparing stars of the *same age* (and therefore of analogous structure), and also if we compare stars of the *same* (homogeneous) *composition.*

We can also try to 'test' the three principal results obtained in Chapter II (Section 7) numerically, for 'homologous' stars of the same type:

$$M'(r') = \text{universal function}\ , \tag{19}$$

$$P_{\text{center}} = \left(\frac{M^2}{R^4}\right)\cdot\text{const}\,, \tag{20}$$

$$T_{\text{center}} = \left(\frac{M}{R}\,\mu_c\right)\cdot\text{const}\,. \tag{21}$$

Instead of making the hypothesis that μ is a universal parameter, we have preferred in these equations to write explicitly its value μ_c at the center of the star in question.

Then, using the values given in Table XXII (for 'massive' stars), we find Table XXVI. We see, as we might have expected, that Equation (20) gives only an order of magnitude, but – and this is what counts – this order of magnitude is quite correct.

Proceeding in the same way to test (21), we derive Table XXVII from Table XXII. This time the test is even more satisfactory than for Table XXIII.

TABLE XXVI

Composition	$X_0=0.70$, $Z_0=0.03$		$X_0=0.90$, $Z_0=0.01$	
Mass	2.82	7.08	2.5	10.0
R (initial)	1.96	3.3	1.59	3.63
M^2/R^4	0.52	0.42	0.98	0.58
P (center) $\times 10^{-16}$	11.8	4.54	14.6	3.4
$10^{-16} P_c/(M^2/R^4)$	22.7	10.8	14.8	5.9

TABLE XXVII

Composition	$X_0=0.70$, $Z_0=0.03$		$X_0=0.90$, $Z_0=0.01$	
Mass	2.82	7.08	2.5	10
$\mu_0=\mu_c$	0.61	0.61	0.53	0.53
R (initial)	1.96	3.3	1.59	3.63
T_6 (center)	23	28	20	28
$M\mu_c/R$	0.88	1.31	0.83	1.46
$T_6/(M\mu_c/R)$	26	21	24	19

In Chapter II we were not able to carry the analysis of 'homologous' models beyond the stage just reviewed, because we had not yet introduced Equation (15) giving $\mathrm{d}T/\mathrm{d}r$ and Equation (16) giving $\mathrm{d}L_r/\mathrm{d}r$.

Now, when $\bar{k}=k_0\varrho^2T^{-3.5}$ and $\varepsilon=\varepsilon_0\varrho T^\nu$, as we have assumed in establishing Equations (15) and (16) at the beginning of Section 4 of the present chapter, we find that Equation (12) gives (since $\varrho\sim MR^{-3}$):

$$T\sim\mu MR^{-1}, \tag{12'}$$

where the sign $\sim$ means equality to within a 'universal function' of r'.
Similarly, making use of (12′) and of the fact that

$$\frac{\mathrm{d}T}{\mathrm{d}r}\sim\frac{\mathrm{d}T}{\mathrm{d}r'}R^{-1}\sim\mu MR^{-2},$$

we find that Equation (15) gives

$$L\sim L_r\sim k_0^{-1}\mu^{7.5}R^{-0.5}M^{5.5}. \tag{15'}$$

Finally, in view of the fact that differentiation only changes the 'universal function' of r', (16) gives:

$$L\sim L_r\sim\frac{\mathrm{d}L_r}{\mathrm{d}r'}\sim\frac{\mathrm{d}L_r}{\mathrm{d}r}R,$$

$$L\sim L_r\sim\varepsilon_0^{+1}\mu^\nu R^{-(3+\nu)}M^{(2+\nu)}. \tag{16'}$$

Equating the expressions for L given by (15′) and (16′), we find at once:

$$R\sim\left[\varepsilon_0k_0\mu^{\nu-7.5}M^{\nu-3.5}\right]^{1/(\nu+2.5)}. \tag{22}$$

Substituting (22) into (15′) *or* (16′), we obtain the final result:

$$L \sim [k_0^{-(\nu+3)} \varepsilon_0^{-1/2} \mu^{(7\nu+22.5)} M^{(5\nu+15.5)}]^{1/(\nu+2.5)} . \tag{23}$$

For 'massive' stars whose energy output is governed by the C-N cycle, we find (taking $\nu = 17$, as in Sears and Brownlee [4], p. 625):

$$\boxed{R \sim (\varepsilon_0 k_0)^{0.05} \mu^{0.49} M^{0.69}} , \tag{22′}$$

and

$$\boxed{L \sim k_0^{-1.02} \varepsilon_0^{-0.02} \mu^{7.3} M^{5.2}} . \tag{23′}$$

Equation (22) is (incorrectly) called the '*mass-radius*' relation and Equation (23) is (incorrectly) called the '*mass-luminosity relation*' – for both of them introduce other parameters besides the mass.

However, ε_0 appears in Equations (22′) and (23′) with an exponent nearly equal to zero, and therefore gives a factor of about 1. Moreover, we know that μ varies within rather narrow limits.

When we compare stars in which k_0 and μ have given values, we should find that the quotients

$$\alpha = \log L/\log M \quad \text{and} \quad \alpha' = \log R/\log M \tag{24}$$

are constant and equal to $\alpha = 5.2$ and $\alpha' = 0.69$, for 'massive' stars.

Table XXI enables us to 'test' Equations (22′) and (23′), insofar as the constancy of α and α' is concerned.

As we might have expected, Table XXVIII shows that α and α' vary with X and Z. For a given X, they also vary with M (for we must recall that the hypothesis of 'homology' is only an approximation). However, the order of magnitude of α and α' is in agreement with Equations (22′) and (23′), which is very encouraging.

TABLE XXVIII

Mass	$X_0 = 0.60$, $Z_0 = 0.02$		$X_0 = 0.70$, $Z_0 = 0.03$	
1.78	$\alpha = 5.3$	$\alpha' = 0.64$	$\alpha = 4.2$	$\alpha' = 0.76$
2.82	$\alpha = 4.6$	$\alpha' = 0.60$	$\alpha = 4.0$	$\alpha' = 0.64$
4.47	$\alpha = 4.3$	$\alpha' = 0.60$	$\alpha = 4.0$	$\alpha' = 0.63$
7.08	$\alpha = 4.1$	$\alpha' = 0.60$	$\alpha = 3.9$	$\alpha' = 0.61$

CONCLUSION

In the preceding study, we have tried to demonstrate three essential aspects of the theory of the internal structure of stars:

(1) the *complexity of the problem*, in spite of many simplifying assumptions;

(2) the fact that, despite this complexity, one can go very far in the physical-numerical analysis of the problem, thanks to

(a) the use of electronic computers,

(b) the advanced state of the theory of radiative transfer,

(c) the relatively satisfactory state of our knowledge of nuclear reactions;

(3) the *danger of intuitively 'obvious' conclusions*, and the necessity of replacing them with physically explicit approximations.

In conclusion, we wish to place particular emphasis on this third, especially instructive, aspect of our study.

One of the principal results of our analysis, which could not have been reached by intuition, is that the *luminosity of a star depends much more strongly on its mass than on the energy parameter* ε.

This results from the fact that the 'physical origin' of L_r, and consequently of L, is to be found in the physically related equations

$$\pi F_\nu = -\frac{4\pi}{3k_\nu}\,\mathbf{grad}_P\,\bar{I}_\nu\,(P) \tag{1}$$

[cf. [9], Chapter VII, Equation (17)], and

$$L_r = -\,4\pi r^2\left(\tfrac{4}{3}\pi\,\frac{\mathrm{d}T}{\mathrm{d}r}\right)\left(\frac{ac}{\pi}\,T^3\,\frac{1}{\bar{k}}\right) \tag{2}$$

[cf. [9], Chapter VII, Equation (51)], which is derived from the theory of radiative transfer in a medium with a certain 'opacity' $\bar{k}$.

Now, the expression for L_r depends essentially on the temperature gradient, which is physically (cf. $P \propto (\varrho/\mu)T$) a consequence of the pressure gradient. And the latter is in turn controlled by the 'struggle' of the star against gravitational collapse.

This set of equations, in which ε does not appear, makes the luminosity depend almost uniquely on the mass. This is why Eddington was able to discover the 'mass-luminosity relation' well before the discovery of the relations which give ε as a function of T.

If this set of equations, which expresses the role of gravity, radiation, and opacity, is completed by the equation

$$\mathrm{d}L_r/\mathrm{d}r = 4\pi r^2 \varrho\varepsilon\,, \tag{3}$$

which introduces ε, the law of variation of ε with T (cf. Chapter IV, Equation (99), with $\nu = 17$) can be used, for example, to derive the following relation for a 'massive' star (cf. Chapter V, Equation (23')):

$$L \sim k_0^{-1.02} \varepsilon_0^{-0.02} \mu^{7.3} M^{5.2}\,. \tag{4}$$

But since L is already almost completely determined – independently of ε – by the mass M alone, any change in one of the factors in Equation (4) requires a compensating change in the other factors. The small *negative* exponent of ε_0 in (4) shows that, paradoxically, a *decrease* in ε_0 by a considerable factor would cause only a small *increase* in L. To be sure, Equation (4) is applicable only to 'homologous' models, but the study of more refined and 'realistic' models reveals the same effect to an even more pronounced degree. Thus in certain cases (cf. [4], p. 582), a *decrease* in ε_0 by a factor of 100 has the result of *increasing* the central density and central temperature, as well as the temperature and pressure gradients throughout the star – the total effect being an *increase* of 13% in the luminosity. Thus a change in ε *does* affect the equilibrium structure and luminosity of the star, but to a smaller extent and, above all, in the opposite sense than would have been suggested by 'common sense' and hasty 'intuition'.

Similarly, when the proportion X of the nuclear fuel (hydrogen) *decreases* in the course of evolution, Equation (4) shows once again that L paradoxically *increases*. Indeed, a decrease in X entails an *increase* in μ (cf. Chapter II, Equation (37)), which appears in (4) with an exponent of $+7.3$. Moreover, since k_0 is proportional to $(1+X)^{0.75}$ (cf. the footnote at the beginning of Section 4, Chapter V), the decrease in k_0 occasioned by a decrease in X also contributes, because of the exponent (-1.02) of k_0 in (4), to an *increase* in L when X decreases.

Thus everything contributes to an increase in L during evolution, and this is in agreement with the data of Table XXV.

The insufficiency of intuitive arguments is particularly evident when we examine the evolution of the *radius* R. Let us consider once more the example of the 'massive' stars, for which we found, under the hypothesis of homology (cf. Section 6 of Chapter V, Equation (22')):

$$R \sim (\varepsilon_0 k_0)^{0.05} \mu^{0.49} M^{0.69}\,. \tag{5}$$

This relation shows that the decrease in X during evolution, which causes μ to increase, should produce a slow *increase* in the radius R (expansion of the star). This is confirmed, for more refined models, by Table XXV.

But a still more detailed study shows that evolution also produces the *contraction* of a sort of 'pseudo-core' which – although it does not coincide exactly with the convective core – can be defined as that part of the mass M, located around the center, whose distance from the center decreases during evolution.

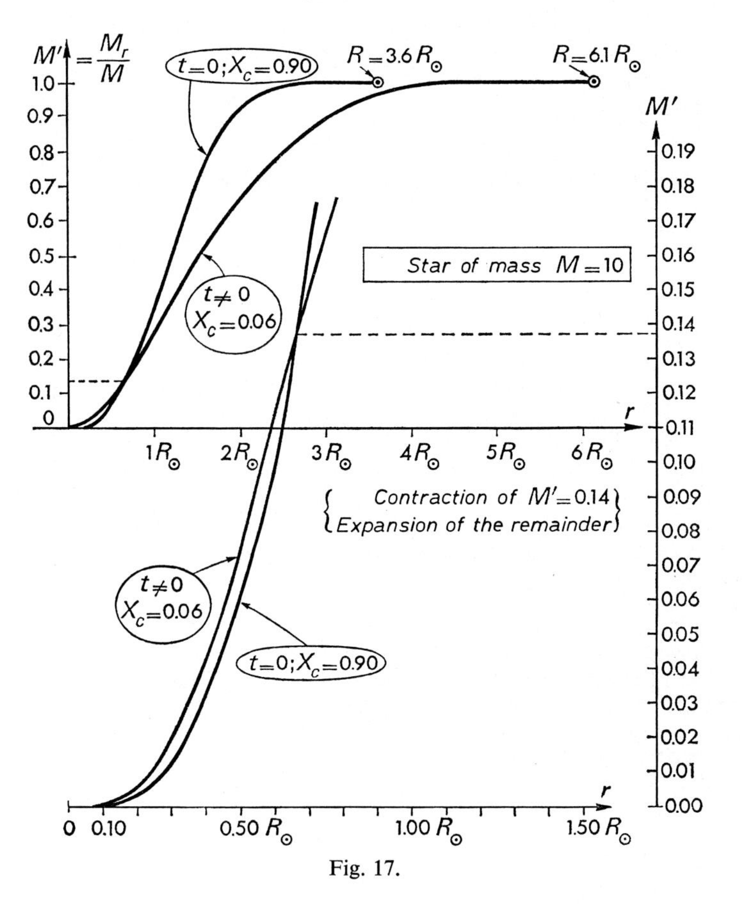

Fig. 17.

This is illustrated in the upper part of Figure 17 (after Schwarzschild, [1], pp. 254 and 258, Tables 28.1 and 28.5), which presents the mass distribution in a 10 $M_{\odot}$ star, both at $t=0$ when $X_c=X_0=0.90$, and at $t_1 \neq 0$ when X_c has been reduced to 0.06.

The lower part of Figure 17 gives these curves in more detail, with expanded scales for r and M'. (The expanded scale for M' is marked at the right.)

Note that in both parts of Figure 17, the abscissa is not $r'=r/R$, but r in units of $R_{\odot}$.

Since the curves 'cross' for M' around 0.14, we see that the corresponding fraction of the total mass does indeed undergo a contraction between the time $t=0$ and the time t_1 (while the 'overall' radius R goes from 3.6 to 6.1 solar radii).

As we see, the theory of the internal structure of stars affords us the pleasure of understanding a great number of peculiarities in the 'life' of a normal star. It gives us a glimpse of the fascination of studying the 'birth' and the 'death' of such stars, to say nothing of the innumerable 'freaks' that populate the stellar universe (magnetic stars, nuclei of galaxies, quasars, pulsars, etc.) which we have not been able to consider in this introductory text.

SOLUTIONS FOR THE EXERCISES

Exercise 1

(1) $u_{\text{rad}} = 10^2$ erg.cm^{-3}; $u_{\text{kin}} = 8 \times 10^{-5}$ erg.cm^{-3}; no.

(2) There is no pressure gradient to oppose the gravitational attraction, and (S) must therefore collapse.

$$\tau = \pi R_0{}^{3/2}/2\sqrt{(2GM)} \approx 10^4 \text{ years},$$

and is thus very short in comparison with the age of the universe. If the density is greater at the center, for example, some pressure gradient will oppose the gravitational contraction and τ will then be greater. (*N.B.* – The symbol $\approx$ means 'approximately equal to'.)

(3a) 1.3×10^{13} erg.g^{-1}.

(3b) $(kT/m_{\text{H}}) \log(\varrho_1/\varrho_0)$.

(3c) $0.8 \times 10^{12} \log(\varrho_1/\varrho_0)$ erg.g^{-1}; $\varrho_1 = 6 \times 10^{-10}$ g.cm^3; $u_{\text{rad}} \approx 1.7 \times 10^{11}$ erg.g^{-1} is then roughly of the same order of magnitude as u_{kin}, since $u_{\text{kin}} \approx 1.3 \times 10^{13}$ erg.g^{-1}.

(4) The order of magnitude of N_2 can be found by applying Boltzmann's law, even though the condition of thermodynamic equilibrium is not fulfilled at the time t_0. Thus: $N_2 \approx 4 \times 10^2$ cm^{-3}.

(5) Since N_1 is much greater than N_2, a photon φ_1 will undergo many interactions on its outward path with atoms in the ground state.

(6) $L_2 \approx 2.5 \times 10^{14}$ cm.

Exercise 2

(1.1) $\varrho_s = \varrho_{00}(1 - \varepsilon)$. (1)

(1.2) $\mathrm{d}P/\mathrm{d}y = - \varrho_s g$. (2)

(2) $\mathrm{d}P'/\mathrm{d}y' = - (1 - \varepsilon)$. (2′)

From (2′) we deduce that $\mathrm{d}y' = -\mathrm{d}P'/(1-\varepsilon)$, whence we can form a table giving the values of $\Delta y'$ associated with the table of values of $-\Delta P'/(1-\varepsilon)$ for intervals $\Delta\varepsilon$; and since $y' = 0$ for $\varepsilon = 0$, we can draw up a table of values of y' as a function of ε.

(3) $P_{0,M} = 0.165(P_{0,E} - P_{A,E})$.

(4.1) $(\mathrm{d}/\mathrm{d}P')\,[\varrho_g(P') \cdot v(P') \cdot \varepsilon(P')]\,\mathrm{d}P' = \varrho_s q\,\mathrm{d}y$. (3)

(4.2) $\varepsilon v(P') = (q/g)\,(RT/\mu)\,[(1/P') - 1]$.

(4.3) For the Earth: $T = 1120$, $g = 980$, whence we obtain Equation (4).

(4.4) For the Moon: $\varepsilon v = 11.3[(1/P') - 1]$. (5)

(5.1) and (5.2) $P' = (1-\varepsilon)/[1 - \varepsilon + 8\varepsilon^3 g^2 \varrho_{00}\mu D^2/(qRT)]$.

(5.3) $F = 8g^2\varrho_{00}\mu D^2/qRT$. For the Earth: $F \approx 1$; for the Moon: $F \approx 2.6 \times 10^{-2}$.

(6.1) See Table A.

(6.2) $P_{0,E} \approx 3.6 \times 10^6$ c.g.s.

(6.3) $y(0.8) \approx 30$ m for the Earth; $y(0.8) \approx 2.4$ m for the Moon.

(6.4) $P_M(0.8) \approx 0.4 \times 10^6$ c.g.s.

(7.1) The height $y(0.8) = 2.4$ m for the Moon is relatively small in comparison with the lunar relief; we may therefore suppose that a value $\varepsilon > 0.8$ will always be attained, and that the dust will thus be carried outside the chimney by the gas flow.

(7.2) Meteoritic origin.

TABLE A

ε	P'_E	$\Delta y'$	y'_E	P'_M	$\Delta y'$	y'_M
0.00	1.000		0	1.000		0
		1.05×10^{-3}			0.000	
0.10	0.999		1.05×10^{-3}	1.000		0
		10.6×10^{-3}			0.000	
0.20	0.990		11.65×10^{-3}	1.000		0
		37.4×10^{-3}			1.33×10^{-3}	
0.30	0.962		49.05×10^{-3}	0.999		0.001 33
		89.2×10^{-3}			3.08×10^{-3}	
0.40	0.904		0.138	0.997		0.004 41
		0.191			7.27×10^{-3}	
0.50	0.799		0.329	0.993		0.011 68
		0.334			17.8×10^{-3}	
0.60	0.649		0.663	0.985		0.029 48
		0.187			13.2×10^{-3}	
0.64	0.578		0.850	0.980		0.042 68
		0.218			17.6×10^{-3}	
0.68	0.504		1.068	0.974		0.060 28
		0.250			30×10^{-3}	
0.72	0.429		1.318	0.965		0.090 28
		0.292			50×10^{-3}	
0.76	0.353		1.610	0.952		0.140 28
		0.327			81.8×10^{-3}	
0.80	0.281		1.937	0.934		0.222 08

Exercise 3

(1) If $\mathbf{g}$ is the gravitational field, $\mathbf{g} = -\mathbf{grad}\,\varphi(r)$; $\varphi(R) = -GM/R$ for a mass M distributed in spherical homogeneous layers, of radius R.

(2) $v_e(R) = [2GM/R]^{1/2}$.

(3) $P_0 = \frac{1}{3}\varphi_0\bar{\varrho}$; $T_0 = \eta\varphi_0$.

(4) $M' = r'^3$; $P' = \frac{3}{2}(1 - r'^2)$.

(5) $P_0 = \frac{2}{3}P_c{}^h$; $T_0 = 2T_c{}^h$.

Exercise 4

(1a) The graphical representation of $\log F$ as a function of r' shows that the representative points are rather well distributed along a straight line of negative slope: $\log F = Ar' + B$, where A is negative. This is in agreement with Equations (1), (2), (3), and (4).

(1b) Computing A and B by the method of least squares, we find: $\alpha = 14.87$; $P'_c = 993 \times 10^{15}$; $\beta = 11.57$; $\varrho'_c = 403$; $\tau = 3.31$; $T'_c = 18 \times 10^6$; $\tau' = 2.945$; $A_0 = 16.11$.

(1c) $\tau \approx \tau'$; $\tau \approx \alpha - \beta$.

(2) $\mu = RT'_c\varrho'_c/P'_c \approx 0.61$, a very reasonable value.

(3) $A_0 = \alpha P'_c/1.91 \times 10^{15}\,\varrho'_c$.

(4) $A_0 \approx 19.21$; $\alpha - \beta$ is not strictly equal to τ.

(5) $r' = 1.0$; $A_0 r'^2 \exp(-\tau' r') = 0.70$; $r' = 0.7$; ... $= 0.94$; $r' = 0.3$; ... $= 0.64$. This result is in good agreement with M'_{exact} for $r' = 0.3$ and $r' = 0.7$.

(6) $M'/r'^2 \to 0$ when $r' \to 0$, and $A_0 \exp(-\tau' r')$ does not approach zero.

(7) $(M')^2 = (C/\alpha^4)\,\{1 - \exp(-\alpha r')\,[1 + \alpha r' + \alpha^2 r'^2/2 + \alpha^3 r'^3/6 + \alpha^4 r'^4/24]\}$.

(8) $\varphi(14.87) \approx 2712$.

(9) $P'_c \approx 914 \times 10^{15}$, in good agreement with the result of question (1).

(10)

r'	0.3	0.4	0.5	0.6	0.7	0.8	0.9	1.0
$\varphi(\alpha r')$	466	112	232	428	741	1201	1935	2712
$10^3 \times M'$	680	840	930	970	988	996	998	1000

(11) The empirical representation of question (1) cannot be simply reduced to the formula found in question (7). These are, in fact, two different approximate representations of M'.

Exercise 5

(1.1) $T = \text{const}$: a rough approximation in the neighbourhood of the surface.

$\mu = \text{const}$: a reasonable hypothesis for r' greater than 0.3; a rough approximation for r' less than or equal to 0.3.

$P = P'_c \exp(-\alpha r')$: a reasonable hypothesis for r' between 0.2 and 0.8; elsewhere, a rough approximation.

(1.2) $\varrho = \varrho'_c \exp(-\alpha r')$.

$$M' = (8\pi R^3/M)\,\varrho'_c \alpha^{-3}\{1 - \exp(-\alpha r)\,[1 + \alpha r' + \alpha^2 r'^2/2]\}. \quad (2)$$

(1.3) $$M'_{r'=1} \approx 4.26\,\varrho_c'/\alpha^3. \quad (3)$$

(1.4) $\alpha^3 \approx 4.26\,\varrho'_c$.

(1.5) $\varrho'_c \approx 746$ c.g.s.

(2.1) $\varrho'_c = (\mu/RT)\,P'_c$, and we have adopted for T a mean value smaller than the central temperature, while for P'_c we have adopted a value greater than the central pressure; hence $\varrho'_c \approx 8\,\varrho_c$.

(2.2) From $T = T'_c \exp(-\tau r')$ we deduce $\varrho'' = (\mu P'_c/RT'_c)\exp-(\alpha-\tau)r'$ and $\varrho''_c \approx 350$.

(2.3) ϱ''_c is too high because the laws adopted for the derivation of $P(r')$ and $T(r')$ are not applicable near the center. The value of P'_c is overestimated.

Exercise 6

(I.1) $P_g = (k/H)(\varrho_g T/\mu)$; $P_r = \frac{1}{3}aT^4$; $P = K(\beta)\varrho_g{}^{4/3}$.

(I.2) $E = \varrho_g[c^2 + K(\beta)\varrho_g{}^{1/3}(3 - \frac{3}{2}\beta)]$.

(II.1) $\mathrm{d}M(r)/\mathrm{d}r = 4\pi r^2 \varrho_g$; $\mathrm{d}P/\mathrm{d}r = -[GM(r)/r^2]\varrho_g$.

The theory of relativity.

(II.2) $$P = \alpha c^2 \varrho_{gc}\theta^4 \quad (1')$$
$$E = \varrho_{gc}c^2\theta^3[1 + \alpha\theta(f-1)]. \quad (2')$$

(II.3) $\theta(0) = 1$; $v(0) = 0$.

$$\mathrm{d}v/\mathrm{d}\xi = \xi^2\theta^3[1 + \alpha\theta(f-1)] \quad (\mathrm{I}')$$

$$\lambda^2 = \alpha c^2/\pi G\varrho_{gc}. \quad (\mathrm{IV})$$

(I′) and (II′) form a system of two first-order differential equations, where the unknown functions are $\theta(\xi)$ and $v(\xi)$ and the boundary conditions are $\theta(0) = 1$ and $v(0) = 0$. The solution depends on α and β. ξ_1 corresponds to the surface of the star; ξ_1 and $v(\xi_1)$ depend on α and β.

(III.1) $T/T_c = \theta$; $\varrho/\varrho_c = \theta^4$.

(III.2) $\varrho/\varrho_c = \theta^3[1 + \alpha\theta(f-1)]/[1 + \alpha(f-1)]$; $\varrho_g/\varrho_c = \theta^3/[1 + \alpha(f-1)]$.

(III.3) $M = 9[(1-\beta)^{1/2}/\beta^2\mu^2]\,v(\xi_1)\,M_\odot$.

(III.4) $T_c = 1.08 \times 10^{13}\,\alpha\beta\mu$ (in K).

$\varrho_{gc} = 3.83 \times 10^{16}\,\alpha^3\beta^4\mu^4/(1-\beta)$ c.g.s.

$P_c = 3.44 \times 10^{37}\,\alpha^4\beta^4\mu^4/(1-\beta)$ c.g.s.

(III.5) $\bar{R} = 4.8 \times 10^{-6}[(1-\beta)^{1/2}/\alpha\beta^2\mu^2]\,\xi_1 R_\odot$.

(IV.1) The radiation pressure is dominant.

(IV.2) $$\mathrm{d}v/\mathrm{d}\xi = \xi^2\theta^3(1 + 3\alpha\theta). \quad (\mathrm{I}'')$$
$$[(1 - 8\alpha v/\xi)/(1 + 4\alpha\theta)]\,\xi^2\,\mathrm{d}\theta/\mathrm{d}\xi + v + \alpha\xi^3\theta^4 = 0. \quad (\mathrm{II}'')$$

ξ_1 and $v(\xi_1)$ do not depend on β.

(IV.3) A value of β much less than 1 corresponds to stars much more massive than the Sun.
(IV.4) $M = (9/\beta^2\mu^2)\,v(\xi_1)\,M_\odot$; $\bar{R} = 4.8 \times 10^{-6}\,\xi_1/\alpha\beta^2\mu^2$.
The curves $M = \text{const}$ are horizontal straight lines.
The curves $\bar{R} = \text{const}$ are straight lines of slope $(-\frac{1}{2})$.
(IV.5) $M = \text{const}$: horizontal straight lines; $\bar{R} = \text{const}$: straight lines of slope $(-\frac{1}{2})$, for in the region under consideration β is much less than 1.
(IV.6) $\bar{R} \approx 6.7 \times 10^{-6}\,\mu M^{3/2}$; $\varrho_{gc} \approx 2.5 \times 10^{17}/\mu^3 M^{7/2}$ c.g.s.

$M/M_\odot$	10^6	10^7	10^8	10^9
$\bar{R}/R_\odot$	3.3×10^3	1.0×10^5	3.3×10^6	1.0×10^8
ϱ_{gc}(c.g.s.)	2.0×10^{-3}	6.4×10^{-7}	2.0×10^{-10}	6.6×10^{-14}

(7) Quasars.

Exercise 7

(1) $\mathrm{d}P/\mathrm{d}r = -GM_r\varrho(r)/r^2$. (2)
(2) $\mathrm{d}M_r = 4\pi r^2\varrho(r)\,\mathrm{d}r$. (3)
(3) $(\mathrm{d}/\mathrm{d}r)\{[r^2/G\varrho(r)][\mathrm{d}P(r)/\mathrm{d}r]\} = -4\pi r^2\varrho(r)$. (4)

When $P(r)$ is replaced by $k\varrho(r)$, (4) becomes a differential equation with only one unknown function: $\varrho(r)$.

(4a) $\xi(r=0) = 0$; $\theta(r=0) = 1$; $(\mathrm{d}\theta/\mathrm{d}\xi)_{r=0} = 0$.
(4b) $(1/\xi^2)(\mathrm{d}/\mathrm{d}\xi)(\xi^2\,\mathrm{d}\theta/\mathrm{d}\xi) = -\theta^n$. (4′)

(5a) The model considered here has an index of $n = 5$; for the Eddington model, the index n is equal to 3.
(5b) The function $(1 + \frac{1}{3}\xi^2)^{-1/2}$ is a solution of (4′) and satisfies the boundary conditions. The function $\theta(\xi)$ is monotonically decreasing and approaches zero when ξ approaches infinity.
(6a) $\theta = 0$ corresponds to a density ϱ equal to zero. But $\theta(\xi)$ is never equal to zero for $n = 5$, so long as ξ has a finite value.

(6b) $\xi_s{}^4 = (8\sqrt{3}/3)k_0\varrho_0\alpha$.
(7) $T_s = T_0\theta_s$.
(8) $R = \lambda k^{5/8}\varrho_0{}^{-1/4}$; $\lambda = \lambda_0 k_0{}^{1/4}\pi^{-5/8}G^{-5/8}$.
(9) $M = (9\sqrt{2}/\sqrt{\pi})\varrho_0{}^{-1/5}k^{3/2}G^{-3/2}$.
(10) $L = \nu R_g{}^2\varrho_0{}^{1/5}k^{7/2}$, with $\nu = \frac{1}{3}2^{9/2}\pi^{3/2}\mu^4 R_g{}^{-4}k_0{}^{-1}\sigma G^{1/2}$.
(11a) $A = \nu\xi^{-4}\lambda^{+4} = (3^{-4}/4)\pi\sigma R_g{}^{-4}G^4\mu^{-4}$; $p = -2$; $q = +4$.
(11b) There is a single (L, M, R) relation only for stars having the same value of μ.
(12a) $A \approx 0.26 \times 10^{-67}$ c.g.s.
(12b) $A_\odot \approx 1.1 \times 10^{-78}$ c.g.s.

The model under consideration is not suitable for the Sun.

(13a) Theoretically, we find: $L \approx M^{5/2}$.
(13b) The polytropic model of index 5 and all the simplifying hypotheses: k_0 constant, μ constant, the poor definition of the surface of the star, the neglect of the equation of radiative transfer of energy, etc.

Excercise 8

(1a) $-r^2(4acT^3/3\bar{k})\,(\mathrm{d}T/\mathrm{d}r) = \int_0^r r^2\varrho\varepsilon\,\mathrm{d}r$.
(1b) $C = c =$ (velocity of light); $G = \varepsilon =$ (energy output per unit mass); $k\varrho = \bar{k}$, and k therefore represents the mean absorption coefficient per unit mass.
(2a) $\bar{c} = c$; $\lambda = 1/\bar{k}$; $\varrho C_v = \mathrm{d}u/\mathrm{d}T = 4aT^3$.
(2b) (1) assumes L.T.E. with $\bar{I}_\nu = B_\nu$ and Fick's law (cf. [9], Chapter VII); (2) assumes that all the energy is transported by radiation.
(3a) Here γ represents the constant of gravitation.

(3b) The gravitational force acting on unit volume at a point P results from the gravitational attraction of all the regions of the star and not only from the attraction of the column of gas located above P.

(3c) Here $\bar{G}$ is the mean energy output per unit mass in a sphere of radius r; this quantity depends on r.

(3d) $\mathrm{d}P/\mathrm{d}r = (4\pi c G\varrho/\bar{k}\bar{\varepsilon})\,(\mathrm{d}P_{\mathrm{rad}}/\mathrm{d}r)$.

(4a) The parameter k depends essentially on the temperature T, the composition of the nuclear mixture X, and the density ϱ – thus it depends on r.

(4b) If $k\bar{G} = \text{const}$, $\varrho = \varrho_0 T^3$, where $\varrho_0 = \text{const}$. If in addition μ is constant with respect to r, the value of the corresponding index n is 3.

(4c) R = Boltzmann's constant; m = mass of a hydrogen atom.

(5a) $\mu(1+f) = (1/m_{\mathrm{H}})\,(\varrho/N_e)$.

(5b) $\mu^{-1}(1+f)^{-1} = \frac{1}{2}(X+1)$.

(5c) Yes.

(6) Yes.

Exercise 9

(1)
$$\mathrm{d}P/\mathrm{d}r = -(GM/r^2)\varrho(r), \qquad \text{(II)}$$
$$\mathrm{d}L_r/\mathrm{d}r = 0, \qquad \text{(III)}$$
$$\mathrm{d}T/\mathrm{d}r = -(3/4ac)\,(\bar{\kappa}/T^3)\,(L/4\pi r^2). \qquad \text{(IV)}$$

We know only that: (Mass of the envelope$/M) \ll 1$; $P_{\mathrm{rad}}/P \ll 1$.
We know nothing concerning the gradients.
The factor depending on the chemical composition.

(C2) $5\mathrm{d}T/T = 2\mathrm{d}P/P$;
$$T = BP^{2/5}. \qquad (1)$$

(C3)
$$\mathrm{d}P/\mathrm{d}r = -(GM/r^2)\,(H\mu/Bk)P^{3/5};$$
$$P^{2/5} = \tfrac{2}{5}(GMH\mu/Bk)[(1/r)-(1/R)]. \qquad (2)$$
$$T = \tfrac{2}{5}(GMH\mu/k)\,[(1/r)-(1/R)] \quad (B \text{ has disappeared!}). \qquad (3)$$

(C4) $T = 5.44 \times 10^6 z'$.

r'	0.86	0.90	0.94	0.98	1.00
T_6	0.89	0.61	0.35	0.11	0.00

(C5) Yes, for $N = 0$, 1, 2, and 3.

(R6) $\mathrm{d}P/\mathrm{d}T = A'T^{7.5}/P$; $A' = (16\pi/3)\,[GkacM/H\kappa_0 L\mu(1+X)]$; $P = AT^{4.25}$ with $A = (A'/4.25)^{1/2}$.

(R7) $T_{\mathrm{rad}}(r') = (1/4.25)\,(H/k)\,(GM/R)\mu z'$.

(R8) $T_{\mathrm{rad}}/T_{\mathrm{conv}} = 0.59$.

(R10) In principle, there should always be an outer convective zone.

(R11a) $y = P_{\mathrm{gas}}/P_{\mathrm{rad}}$.

(R11b) $\mathrm{d}(y^2)/\mathrm{d}x = 16[1 - y(y+1)/x]$.

(R11c) Yes, for $y \gg 1$.

(R11d) $y^2 = (16/17)\,x$.

(R11e) We set $y = f(x) + g(x)$, where $f(x)$, the solution of (c), is much greater than the absolute value of $g(x)$. In the first approximation, we find $g = -16/33$.

Exercise 10

(1) $R \approx 6\,R_\odot$; $M \approx 10\,M_\odot$. The central convective zone, in contrast with the outer convective zone of the Sun. The homogeneous central composition.

(2) $y(r')$ should be constant for r' between 0.00 and 0.06. This is true for the values of P and T in Table (T).

(3) *Zone* (S), r' greater than or equal to 0.16:
Extremely small output of nuclear energy, for ϱX^2 is small, whence $L' \approx 1$.
Homogeneous composition, for the rate of nuclear reactions is almost zero.

Radiative equilibrium, for the temperature gradient cannot become high enough for convection.

Zone (C), r' between 0 and 0.0786:

Luminosity different from 1, for thermonuclear reactions take place here.

Convective equilibrium, for the temperature gradient is too high for radiative equilibrium. Homogeneous composition, thanks to convective mixing.

Zone (I), r' between 0.08 and 0.16:

Constant luminosity, for the output of nuclear energy is very small (ϱX and ϱX^2 are small).

X and Y vary with r' because of thermonuclear reactions.

(4) $\mathrm{d}\log T/\mathrm{d}r' = -5.33 \times 10^{27}\, \bar{k}L'/T^4 r'^2$; Table ($T$) is accurate for r' between 0.40 and 0.42, and inaccurate for r' between 0.08 and 0.10.

(5) $\mathrm{d}L'/\mathrm{d}r' = 1.288 \times 10^{-5} \varrho^2 X r'^2 (T^4 \times 10^{-28})^4$. The computation of $[\Delta L']_{\Delta r'=0.02}$ shows that the agreement with the values of L' in (T) is not very satisfactory.

(6) $P_{\mathrm{gas}} = 8.317 \times 10^{-2} (\varrho/\mu)\,(T \times 10^{-7})$.

$P_{\mathrm{rad}} = 2.520 \times 10^{-3} (T^4 \times 10^{-28})$.

$P \times 10^{-15} = 10(P_{\mathrm{gas}} \times 10^{-16} + P_{\mathrm{rad}} \times 10^{-16})$.

(7) $\Delta(r') \approx 0.34$ for r' between zero and 0.08; $\Delta(r')$ decreases from 0.34 to zero for r' between 0.08 and 0.16, and remains zero for r' greater than or equal to 0.16. Thus we note a systematic effect in the region where r' is less than or equal to 0.08. The variation of $\Delta(r')$ with r' has the same appearance as that of $1/\mu$, and this systematic effect can be attributed to the use of a value of $1/\mu$ that is systematically too high: namely, $1/\mu = 1.87$ (or $X = 0.900$) instead of $1/\mu = 0.824$ in the region in question.

(8) $(\mathrm{d}P/\mathrm{d}r') \times 10^{-17} = -3.129 \times 10^{-5} (M' \times 10^3) \varrho / r'^2$.

We find that the discrepancy between the calculated values of $P \times 10^{-16}$ and the values in Table (T) is less than 0.01 in zone (S); but in the other two zones (C and I), this discrepancy is greater than 0.01 and increases as r' decreases.

(9) $T_c = 33.57 \times 10^6$; $\log T_\varrho = 0.4 \log (P_\varrho/P_c) + 7.526$.

The difference between $\log T$ in Table (T) and $\log T_\varrho$ is equal to 0.019, and remains constant when r' varies.

(10) The difference $\Delta \log T = 0.019$ confirms the fact that the pressure appearing in Table (T) is inaccurate for r' less than or equal to 0.06, while the values of $\log T$ are accurate.

BIBLIOGRAPHY

1. Works Cited in the Text (Principal Sources)

[1] Schwarzschild, M.: 1958, *Structure and Evolution of the Stars*, Princeton University Press, (reprinted in 1965 by Dover Publications, New York).
[2] *Ann. Rev. Astron. Astrophys.* **5**, Palo Alto, Calif., U.S.A., 1967.
[3] Tayler, R. J.: 1967, *Monthly Notices Roy. Astron. Soc.* **135**, 225.
[4] *Stellar Structure*, in *Stars and Stellar Systems*, Vol. VIII (general ed. G. P. Kuiper), University of Chicago Press, 1965.
[5] *Stellar Structure*, in *Encyclopedia of Physics*, Vol. LI, Springer, Berlin, 1958.
[6] Chandrasekhar, S.: 1939, *An Introduction to the Study of Stellar Structure*, University of Chicago Press (reprinted in 1967 by Dover Publications, New York).
[7] Cowling, T. G.: 1935, *Monthly Notices Roy. Astron. Soc.* **96**, 42.
[8] Burbidge, E. M. and G. R., Fowler, W. A., and Hoyle, F.: 1957, *Rev. Mod. Phys.* **29**, 547.
[9] Kourganoff, V.: 1967, *Introduction à la théorie générale du transfert des particules*, Dunod, Paris. English translation: *Introduction to the General Theory of Particle Transfer*, Gordon and Breach, New York, 1967.

N.B. – The references given *below* include only *general* works (books, monographs, review articles), to the exclusion of all the 'original articles' dispersed in the various journals. The indicated 'reviews' include only *recent* publications (1962–71), for all the review articles previous to 1962 have already been 'exploited' in the works cited above, especially in [4] and [5]. References to the original articles can easily be found in Section 0.65 of *Astronomy and Astrophysics Abstracts*, Vol. 1–6....

2. General Works, Books, Review Articles, and Monographs (not Cited)

A. FUNDAMENTAL WORKS

Eddington, A. S.: 1926, *The Internal Constitution of the Stars*. Reprint Dover, 1959.
Jeans, J.: 1928, *Astronomy and Cosmogony*. Reprint Dover, 1961.
Rosseland, S.: 1949, *The Pulsation Theory of Variable Stars*. Reprint Dover, 1964.
Allen, C. W.: 1954, *Nuclear Transformations, Stellar Interiors, and Nebulae*, Ronald.
Schatzman, E.: 1958, *White Dwarfs*, North-Holland, Publ. Co., Amsterdam.
Burbidge, G.: 1962, 'Nuclear Astrophysics', in *Ann. Rev. Nucl. Sci.* **12**, Palo Alto, Calif.
Menzel, D. H.: 1963, *Stellar Interiors*, Chapman Hall, London.
Chiu, H. Y.: 1965, *Neutrino Astrophysics*, Gordon and Breach, New York.
Meadows, A. J.: 1967, *Stellar Evolution*, Pergamon, Oxford, 1967.
Shen, B. S.: 1967, *High Energy Nuclear Reactions in Astrophysics*, Benjamin, New York, 1967.
Page, T.: 1968, *The Evolution of Stars*, Macmillan, London.
Chiu, H. Y.: 1968, *Stellar Physics*, Vol. I, Blaisdell, Waltham, Mass.
Clayton, D. D.: 1968, *Principles of Stellar Evolution and Nucleosynthesis*, McGraw-Hill, New York.
Cox, J. P. and Giuli, R. T.: 1968, *Principles of Stellar Structure*, Vol. I: *Physical Principles*; Vol. 2: *Application to Stars*; Gordon and Breach, New York.
Novotny, E.: 1971, *An Introduction to Stellar Atmospheres and Interiors*, Oxford University Press, New York.

B. REVIEW ARTICLES AND MONOGRAPHS (1962–71)

[References between square brackets are given in Section C below.]

B1. Physical Bases

'Dimensionless Ratios and Stellar Structure', by E. E. Salpeter [10], pp. 463–75.
'Astrophysical Problems', by T. A. Tombrello [11], pp. 195–212.
'Theories of the Equation of State of Matter at High Pressures and Temperatures', by G. Brush Stephen [12], pp. 1–137.
'Symmetries of Strong Interactions', by N. Dallaporta [13], pp. 400–17.
'Properties of Cold, Dense Matter', by G. Szamosi [13], pp. 281–301.

B2. General Reports on the Structure of the Stars

Section on the internal structure of stars, in [14].
'The Outer Regions and the Internal Constitution of the Stars', by P. Ledoux (in Russian) [15], No. 6, pp. 14–20.
'Stellar Constitution', by A. G. Massevitch [16], pp. 405–23.
'Ordinary Stars, White Dwarfs and Neutron Stars', by L. C. Green [44], pp. 18–20.

B3. Construction (Computation) of Stellar Models

'Construction of Stellar Models', by M. H. Wrubel [47], pp. 50–84.
'Stars with He-rich Cores', by W. A. Fowler [17], pp. 330–68.
'Methods for Calculating Stellar Evolution', by R. Kippenhahn *et al.* [18], pp. 129–218.
'Stellar Evolution According to Numerical Models', by A. Weigert [19], pp. 212–82.

B4. Protostars, Star Formation, Prestellar Evolution

'Evolution of Protostars', by Hayashi Chushiro [20], pp. 171–92.
'Protostars', by A. G. W. Cameron [21], pp. 131–45.
'The Birth of Stars', by S. B. Pikel'ner [22], pp. 42–49.
'Sternentstehung und Frühphase der Sternentwicklung', by K. Fritze [23], pp. 64–79.
'Theory of Star Formation', by G. B. Field [24], pp. 29–45.
'Prestellar Evolution', by P. Ledoux [25], pp. 208–35.
'Theories of Star Formation', by D. McNally [26], pp. 71–108.
'Some Problems of Star Formation', by V. C. Reddish [27], pp. 173–94.

B5. Stellar Evolution

'Stellar Stability and Stellar Evolution', by P. Ledoux [17], pp. 394–445.
'Stellar Evolution', by R. J. Tayler [28], Vol. 31, pp. 167–223.
'Stellar Evolution', by E. E. Salpeter [29], pp. 71–88.
'Equation of State and Stellar Evolution', by E. E. Salpeter [30], pp. 205–09.
'Stellar Evolution', by I. Iben (in Russian) [15], pp. 19–26.
'Evolution kosmischer Materie', by A. Unsöld [31].
'Les progrès de la théorie de l'évolution stellaire', by A. Boury and M. Gabriel [32], pp. 27–49.
'Stellar Evolution', by A. G. W. Cameron [25], pp. 236–49.

B6. Energy Production in Stars

'Energy Production in Stars', by H. A. Bethe [33].
'Nuclear Energy Generation in Supermassive Stars', by W. A. Fowler [10], pp. 413–27.

B7. Massive Stars, Superdense Bodies, Gravitational Collapse

'Relativistic Astrophysics and Gravitational Collapse', by S. M. Chitre [48], pp. 307–24.
'Massive Stars, Relativistic Polytropes and Gravitational Radiation', by W. A. Fowler [34], pp 545–55.
'Present State of the Theory of Superdense Bodies', by V. A. Ambarzumian [35], pp. 91–131.
'Superdense Stars', by J. A. Wheeler [20], pp. 393–432.
'Superdense Matter in Stars', by M. Ruderman [36], pp. 152–60.
'Solid Stars', by M. Ruderman [37], pp. 24–31.
'Physics of the Catastrophe Stage of the Compression of a Star', by Y. B. Zel'dovich and I. D. Novikov [38], pp. 66–78.

B8. Nucleosynthesis in Stars

'High-Energy Nucleosynthesis', by H. Reeves [39], pp. 163–226.
'Nucleosynthesis', by C. J. Hansen *et al.* [40].
'La nucléosynthèse des éléments chimiques dans les étoiles', by J. Audouze [41], pp. 81–88, 137–45.
'The Empirical Foundation of Nucleosynthesis', by W. A. Fowler [42], pp. 3–29.
'Nuclear Reactions on Stellar Surfaces and Their Relation to Stellar Evolution', by H. Reeves [19], pp. 283–419.
'Nucleosynthesis: General Report', by J. L. Greenstein [43], pp. 85–91.
'Neutron Capture Data and Stellar Temperatures', by R. L. Macklin [46], pp. 166–76.

B9. Origin of the Elements

'L'abondance des éléments cosmiques et leur origine', by M. Hack [45], pp. 147–53.
'The Origin of the Elements', by D. D. Clayton, [47], pp. 28–36.

B10 Neutron Stars and White Dwarfs

'Present Status of the Neutron Star Hypothesis', by A. Finzi [13], pp. 302–12.
'Neutron Stars', by A. G. W. Cameron [49], pp. 178–208.
'Recent Developments in the Theory of Degenerate Dwarfs', by J. P. Ostriker [51], pp. 323–52.

B11. Stellar Pulsations

'Physical Basis of the Pulsation Theory of Variable Stars', by S. A. Zhevakin [52], pp. 367–400.
'The Calculation of Stellar Pulsations', by R. Christy [34], pp. 555–71.
'Theoretical Models for the Pulsation of Cepheids', by N. Baker [17], pp. 368–88.
'Pulsation Theory', by R. Christy [20], pp. 353–92.
'Oscillation Theories of Magnetic Variable Stars', by P. Ledoux [54], pp. 65–81.
'The Linear Theory Initiation of Pulsational Instability in Stars', by J. P. Cox [55], pp. 3–104.
'The Non-Linear Calculations for Pulsating Stars', by R. F. Christy [55], pp. 105–56.
'Oscillations et Stabilité Stellaires', by P. Ledoux [19], pp. 44–211.

B12. Stellar Rotation

'Rotation and Magnetism in Stellar Structure and Evolution', by I. W. Roxburgh [54], pp. 45–64.
'Rotation and Stellar Interiors', by I. W. Roxburgh [56], pp. 9–19.

B13. Mass Loss in Stars

'Mass Loss in Stars', by R. Weymann [52], pp. 97–144.

B14. Neutrinos in Stars

'Astrophysical Neutrinos', by M. A. Ruderman [28], Vol. 28, pp. 411–62.
'Neutrino Astrophysics', by W. A. Fowler [13], pp. 367–99.
'Neutrino Physics and Neutrino Astrophysics', by P. V. Ramana Murthy [57], pp. 272–91.
'Neutrinos from the Sun', by E. Salpeter [58], pp. 97–102.
'On the Investigation of the Internal Structure of the Sun by Means of Neutrino Radiation Studies', by V. A. Dergachov and G. E. Kocharov [59], pp. 491–93.

B15. Convection in Stars

'Convection in Stars', by E. A. Spiegel [51], pp. 323–52.

C. REFERENCES FROM [10] TO [59].

[10] *Perspectives Mod. Phys.*, Interscience, New York, 1966.
[11] *Nucl. Res. Low Energy Accelerators*, Academic Press, New York, 1967.
[12] *Progr. High Temperat. Phys. Chem.*, Vol. 1, Pergamon, Oxford, 1967.
[13] *Proc. Intern. School Phys. Enrico Fermi*, Vol. 35, Academic Press, New York, 1965.
[14] *Trans. IAU* **13A**, *Reports on Astronomy*, Reidel, Dordrecht, 1967.
[15] *Zemlya i Vselennaya*, No. 4, 1969.
[16] *Trans. IAU* **14A**, Reidel, Dordrecht, 1970.
[17] *Rend. Scuola Intern. Fis. Enrico Fermi*, Vol. 28, Academic Press, New York, 1963.

[18] *Methods Computat. Phys.*, Vol. 7, Academic Press, New York, 1967.
[19] *La Structure Interne des Etoiles*, XIe Cours de perfectionnement de l'Association Vaudoise des Chercheurs en Physique, Observatoire de Genève, Sauverny, Suisse (Saas-Fee), 1970.
[20] *Ann. Rev. Astron. Astrophys.* **4**, Palo Alto, Calif., U.S.A. 1966.
[21] *Infrared Astronomy*, Conference Goddard Space Flight Center, 1966.
[22] *Priroda*, No. 11.70 (1970).
[23] *Sterne* **47** (1971).
[24] *Évolution Stellaire avant la Séquence Principale*, 16e Colloque de Liège, Inst. Astrophys. Liège, 1970.
[25] *Structure and Evolution of the Galaxy*, Proc. NATO Adv. Study Inst., Reidel, Dordrecht, 1971.
[26] *Rep. Progr. Phys. G.B.* **34**, No. 2 (1970).
[27] *Vistas Astron.* **7**, Pergamon, Oxford, 1966.
[28] *Repts. Progr. Phys. Inst. Phys. Soc.*, Vol. 28, 1965; Vol. 31, 1968.
[29] *Structure and Evolution of Galaxies*, Interscience, New York, 1965.
[30] *Low-Luminosity Stars*, Symposium, 1968.
[31] *Sonderdruck Sternw. Kiel*, No. 150 (1968).
[32] *Ciel et Terre* **87** (1971).
[33] *Phys. Today* **21**, No. 9 (1968); see the same article in *Science* **161**, No. 3841 (1968), 541–47.
[34] *Rev. Modern Phys.* **36**, No. 2 (1964).
[35] *Voprosy Cosmogonyi*, **9** (1963), Moscow.
[36] *Physique Fondamentale et Astrophysique*, Colloque du CNRS, 1969, Publ. CNRS, Paris, 1970.
[37] *Scientific American* **224**, No. 2 (1971).
[38] *Problems of Astronomy and Geodesy*, Nauka, Moscow, 1970.
[39] *Hautes énergies en astrophysique*, Vol. 2, New York-London-Paris, 1967.
[40] *Nucleosynthesis*, Proc. Conf. NASA, New York, 1965 (ed. by W. D. Arnett, C. J. Hansen, J. W. Truran, and A. G. W. Cameron), Gordon and Breach, New York, 1968.
[41] *Science Progrès, La Nature* **97** (1969).
[42] *Origin and Distribution of the Elements*, Symposium, Paris, 1967.
[43] *Symposium on Stellar Composition and Nucleosynthesis*, *Comments Astrophys. Space Phys.* **2** (1970).
[44] *Sky Telesc.* **41** (1971).
[45] *Sciencia* **99**, No. 12 (1964).
[46] *Rev. Modern Phys.* **37**, No. 1 (1965).
[47] *Phys. Today* **22**, No. 5 (1969).
[48] *Proc. 10th Symp. Cosmic Rays, Elementary Particle Phys. Astrophys.*, Aligarth, 1967.
[49] *Ann. Rev. Astron. Astrophys.* **8**, Palo Alto, Calif., 1970.
[50] *Supernovae and Their Remnants*, Proc. Conf. on Supernovae, 1967, Gordon and Breach, New York, 1969.
[51] *Ann. Rev. Astron. Astrophys.* **9**, Palo Alto, Calif., 1971.,
[52] *Ann. Rev. Astron. Astrophys.* **1**, Palo Alto, Calif., 1963.
[53] *Trans. IAU* **11A**, Academic Press, New York, 1962, pp. 413–17.
[54] *Magnet. Related Stars*, Mono Book Corp., Baltimore, Md., 1967.
[55] *Fifth Symp. Cosmic Gas Dynamics*, London, 1967, Academic Press, New York.
[56] *Stellar Rotation*, Proc. IAU Colloquium, 1969, Reidel, Dordrecht, 1970.
[57] *Proc. 9th Sympos. Cosmic Rays, Elementary Particle Phys. Astrophys.*, Bombay, 1966.
[58] *Comments Nucl. Particle Phys.* **2**, No. 4 (1968).
[59] *Can. J. Phys.* **46**, No. 10, Part 3 (1968).

INDEX OF SUBJECTS

Abundances X, Y, and Z *24*
Accumulation of errors *26*
Active component *8*
Actual determination of the distributions ϱ (r) and X (r) for each model *85*
Age of a star *97*
Atomic mass (unit of) *47*
Atomic masses of the different nuclei *47*
Atomic weight of the atoms *48*

Binding energy between the nucleus and the electrons *51*
Boltzmann's constant *12*
Boundary conditions *24*

Central pressure 30
Central region *76*
Chemical composition 1, 2
C-N cycle *45*
Collision crosssection (microscopic) *53*
Composition of nuclei (schematic representation) *45*
Constant of gravitation *4*
Contraction (gravitational) *85*
Convective core *91*
Convective region (outer) *92*
Convective zone (superficial) *35*
Convergence of cyclic reactions (to a stationary state) *61*
Convergence (physical mechanism involved in the, towards a stationary state) *66*
Conversion of the mass into energy *2*

Dimensionless variables *26*
Dynamical pressure *8*

Electrical neutrality of the plasma *51*
Emergent flux of the field **g** *4*
Empirical representation (of ε_{pp} and ε_{CN}) *70*
Empirical representation (of the functions $g(r')$, $\varrho(r')$, $P(r')$, and $T(r')$) *32*
Energy ε (calculation of the) *52*
Energy equilibrium (equation of) *44*
Energy produced (average, per second by nuclear reactions in each gram) *44*
Energy sources 2
Errors (accumulated) *28*
Errors (relative, of the approximate calculation) *29*
Evolutionary models *84*
Evolutionary sequences *85*
Evolution of the distributions $X(r)$ and $Y(r)$ *87*
Exact values *26*
Expression for ε_{12} in ergs per gram per second 57

Final test *73*
Force per unit volume (gravitational) 7
Force per unit volume produced by the pressure gradient *8*
Fossilized composition *85*
Fossilized helium *87*
Fusion of two pairs of protons *45*

Gamow energy E_G *53*
Gamow fraction $f_G(W)$ *53*
Gas constant *13*
Gas pressure (P_{gas}, calculation of the) *12*
Gauss's theorem for electrostatics *4*
General mixing *85*
General relations (connecting the functions g, ϱ, P, and T) *34*
Gravitational contraction *2*
Gravitational field *4*
Gravitational force per unit volume *4*
Gravitational interaction *3*
Gravitational potential *11*

Heavy elements *13*
Homologous model *18*, *98*
Homologous stars 91

Initial chemically homogeneous star *85*
Initial composition *85*
Initial phase *85*
Initial phase of the expansion of the Universe *87*
Instability (condition for) *36*
Integrated radiation (net flow of) *44*
Integration methods (principles of the) *93*
Internal structure *24*
Ionization (complete) *13*, 16

Kramers-Rosseland opacity law *84*

Law of attraction by universal gravitation *4*
Layer (convective) *36*
Layers of stars (outer) 2
Local interaction 3
Lower main sequence *90*
Low-mass stars *90*
Luminosity $L(R)$ 24
Luminosity of a star 101
Luminosity (total bolometric) *1*

Mass defect *48*
Mass distribution (approximate calculation of the) *27*
Mass excess *48*
Massive stars *90*
Mass-luminosity relation *100*
Mass M 2
Mass M of the star *1*
Mass (mean, μ of a particle of the mixture) *12*, 16
Mass M (processes capable of modifying the) *86*
Mass M_r (determination of the distribution of the) *25*
Mass M_r located inside the sphere of radius r *4*
Mass - radius relation *100*
Mass (reduced) *56*
Mass (total, of the Sun) *15*
Mathematical structure (of the problem) *84*, *93*
Mean duration (of a cycle) *68*
Mean duration of an isolated reaction (R) *60*
Mean lifetime (of a given nucleus with respect to an isolated reaction (R)) *57*
Mechanical equilibrium *3*, 17
Mechanical equilibrium (equation expressing the) 9
Meridian circulation *85*
Model (of the Sun at constant density) *15*

Neutrino loss *49*
Newton's theorem *4*
Nuclear energy (law of, generation) *84*
Nuclear masses *48*
Nuclear reactions 25
Number density *12*
Number of reactions (R_{12} per cm^3 per second) *52*
Numerical integration *25*

Observational data *1*
Obvious conclusions (danger of intuitively) *101*

Parameter τ_{cycle} *65*
Perfect gas law *12*
Physical conditions 1
Physical meaning of $\tau_p(c)$ *59*
Physical system *47*
Poisson's equation *10*
Polytropic index *34*, 91
Potential barrier *2*, 53
p-p chain *45*, 45
Pressure at the center of the Sun *16*
Pressure distribution (approximate calculation of the) *29*
Pressure gradient 30
Pressure (gradient of the total) *3*
Prestellar state *85*
Pseudo-core *102*

Radiation pressure *15*, 32, 74
Radius (of the star) *1*
Radius (solar) *15*
Reaction rate *52*
Reactive component *8*
Real pressure and temperature distributions in the Sun *25*
Relation between M_r and the density *10*
Relative velocities (distribution of) *55*
Right distribution ϱ (r) *25*
Rotation (stellar) *86*

Stratification in homogeneous spherical layers 6
Structure of the star *1*
Struggle of the star against gravitational collapse *101*
System $C^{12} = 12$ *47*

Temperature (central) 16
Temperature (determination of the distribution of the) *31*
Temperature distribution (approximate calculation of the) *31*
Temperature (optimum) *18*
Temperatures (central, in the initial homogeneous phase) *90*
Thermonuclear reactions *2*, 17
Time factor *1*
Total (number of free electrons per gram) 14
Total (number of particles of all kinds per gram) *14*
Total pressure (P, determination of the distribution of the) *28*
Transition probability P_{ca} per reaction (R) *61*
Trapezoid method *26*
Types of structure (different) *90*

Universal function *19*
Universal quantities *19*
Upper main sequence *90*

Variable M_r (use the, to connect two successive evolutionary stages) *88*

ASTROPHYSICS AND SPACE SCIENCE LIBRARY

Edited by

J. E. Blamont, R. L. F. Boyd, L. Goldberg, C. de Jager, Z. Kopal, G. H. Ludwig, R. Lüst, B. M. McCormac, H. E. Newell, L. I. Sedov, Z. Švestka, and W. de Graaff

1. C. de Jager (ed.), *The Solar Spectrum. Proceedings of the Symposium held at the University of Utrecht, 26–31 August, 1963.* 1965, XIV + 417 pp.
2. J. Ortner and H. Maseland (eds.), *Introduction to Solar Terrestrial Relations. Proceedings of the Summer School in Space Physics held in Alpbach, Austria, July 15–August 10, 1963 and Organized by the European Preparatory Commission for Space Research.* 1965, IX + 506 pp.
3. C. C. Chang and S. S. Huang (eds.), *Proceedings of the Plasma Space Science Symposium, Held at the Catholic University of America, Washington, D.C., June 11–14, 1963.* 1965, IX + 377 pp.
4. Zdeněk Kopal, *An Introduction to the Study of the Moon.* 1966, XII + 464 pp.
5. Billy M. McCormac (ed.), *Radiation Trapped in the Earth's Magnetic Field. Proceedings of the Advanced Study Institute, Held at the Chr. Michelsen Institute, Bergen, Norway, August 16–September 3, 1965.* 1966, XII + 901 pp.
6. A. B. Underhill, *The Eearly Type Stars.* 1966, XIII + 282 pp.
7. Jean Kovalevsky, *Introduction to Celestial Mechanics.* 1967, VIII + 427 pp.
8. Zdeněk Kopal and Constantine L. Goudas (eds.), *Measure of the Moon. Proceedings of the Second International Conference on Selenodesy and Lunar Topography held in the University of Manchester, England, May 30–June 4, 1966.* 1967, XVIII + 479 pp.
9. J. G. Emming (ed.), *Electromagnetic Radiation in Space. Proceedings of the Third ESRO Summer School in Space Physics, held in Alpbach, Austria, from 19 July to 13 August, 1965.* 1968, VIII + 307 pp.
10. R. L. Carovillano, John F. McClay, and Henry R. Radoski (eds.), *Physics of the Magnetosphere. Based upon the Proceedings of the Conference held at Boston College, June 19–28, 1967.* 1968, X + 686 pp.
11. Syun-Ichi Akasofu, *Polar and Magnetospheric Substorms.* 1968, XVIII + 280 pp.
12. Peter M. Millman (ed.), *Meteorite Research. Proceedings of a Symposium on Meteorite Research held in Vienna, Austria, 7–13 August, 1968.* 1969, XV + 941 pp.
13. Margherita Hack (ed.), *Mass Loss from Stars. Proceedings of the Second Trieste Colloquium on Astrophysics, 12–17 September, 1968.* 1969, XII + 345 pp.
14. N. D'Angelo (ed.), *Low-Frequency Waves and Irregularities in the Ionosphere. Proceedings of the 2nd ESRIN-ESLAB Symposium, held in Frascati, Italy, 23–27 September, 1968.* 1969, VII + 218 pp.
15. G. A. Partel (ed.), *Space Engineering. Proceedings of the Second International Conference on Space Engineering, held at the Fondazione Giorgio Cini, Isola di San Giorgio, Venice, Italy, May 7–10, 1969.* 1970, XI + 728 pp.
16. S. Fred Singer (ed.), *Manned Laboratories in Space. Second International Orbital Laboratory Symposium.* 1969, XIII + 133 pp.
17. B. M. McCormac (ed.), *Particles and Fields in the Magnetosphere. Symposium Organized by the Summer Advanced Study Institute, held at the University of California, Santa Barbara, Calif., August 4–15, 1969.* 1970, XI + 450 pp.
18. Jean-Claude Pecker, *Experimental Astronomy.* 1970, X + 105 pp.
19. V. Manno and D. E. Page (eds.), *Intercorrelated Satellite Observations related to Solar Events. Proceedings of the Third ESLAB/ESRIN Symposium held in Noordwijk, The Netherlands, September 16–19, 1969.* 1970, XVI + 627 pp.
20. L. Mansinha, D. E. Smylie and A. E. Beck, *Earthquake Displacement Fields and the Rotation of the Earth. A NATO Advanced Study Institute Conference Organized by the Department of Geophysics, University of Western Ontario, London, Canada, June 22–28, 1969.* 1970, XI + 308 pp.
21. Jean-Claude Pecker, *Space Observatories.* 1970, XI + 120 pp.

22. L. N. Mavridis (ed.), *Structure and Evolution of the Galaxy. Proceedings of the NATO Advanced Study Institute, held in Athens, September 8–19, 1969.* 1971, VII + 312 pp.
23. A. Muller (ed.), *The Magellanic Clouds. A European Southern Observatory Presentation: Principal Prospects, Current Observational and Theoretical Approaches, and Prospects for Future Research. Based on the Symposium on the Magellanic Clouds, held in Santiago de Chile, March 1969, on the Occasion of the Dedication of the European Southern Observatory.* 1971, XII + 189 pp.
24. B. M. McCormac (ed.), *The Radiating Atmosphere. Proceedings of a Symposium Organized by the Summer Advanced Study Institute, held at Queen's University, Kingston, Ontario, August 3–14, 1970.* 1971, XI +455 pp.
25. G. Fiocco (ed.), *Mesospheric Models and Related Experiments. Proceedings of the 4th ESRIN-ESLAB Symposium, held at Frascati, Italy, July 6–10, 1970.* 1971, VIII + 298 pp.
26. I. Atanasijević, *Selected Exercises in Galactic Astronomy.* 1971, XII + 144 pp.
27. C. J. Macris (ed.), *Physics of the Solar Corona. Proceedings of NATO Advanced Study Institute on Physics of the Solar Corona, held at Cavouri-Vouliagmeni, Athens, Greece, 6–17 September 1970.* 1971, XII + 345 pp.
28. F. Delobeau, *The Enivronment of the Earth.* 1971, IX + 113 pp.
29. E. R. Dyer (general ed.), *Solar-Terrestrial Physics/1970. Proceedings of the International Symposium on Solar-Terrestrial Physics, held in Leningrad, U.S.S.R., 12–19 May 1970.* 1972, VIII + 938 pp.
30. V. Manno and J. Ring (eds.), *Infrared Detection Techniques for Space Research, Proceedings of the Fifth ESLAB-ESRIN Symposium, held in Noordwijk, The Netherlands, June 8–11, 1971.* 1972. XII + 344 pp.
31. M. Lecar (ed.), *Gravitational N-Body Problem, Proceedings of IAU Colloquium No. 10, held in Cambridge, England, August 12–15, 1970.* 1972, XI + 441 pp.
32. B. M. McCormac (ed.), *Earth's Magnetospheric Processes. Proceedings of a Symposium Organized by the Summer Advanced Study Institute and Ninth ESRO Summer School, held in Cortina, Italy, August 30–September 10, 1971.* 1972, VIII + 417 pp.
33. Antonín Rükl, *Maps of Lunar Hemispheres.* 1972, V + 24 pp.

SOLE DISTRIBUTORS FOR U.S.A. AND CANADA:

Vols. 2–6, and 8: Gordon & Breach Inc., 150 Fifth Ave., New York, N.Y. 10011

Vols. 7 and 9–28: Springer Verlag New York, Inc., 175 Fifth Ave., New York, N.Y. 10011